THE HISTORY of MEDICINE

THE SCIENTIFIC REVOLUTION AND MEDICINE

1450–1700

THE HISTORY of MEDICINE

THE SCIENTIFIC REVOLUTION AND MEDICINE

1450–1700

KATE KELLY

Facts On File
An imprint of Infobase Publishing

THE SCIENTIFIC REVOLUTION AND MEDICINE: 1450–1700

Facts On File, Inc.
An imprint of Infobase Publishing
132 West 31st Street
New York NY 10001

Library of Congress Cataloging-in-Publication Data
Kelly, Kate, 1950-
The scientific revolution and medicine : 1450–1700 / Kate Kelly.
p. cm. — (The history of medicine)
Includes bibliographical references and index.
ISBN-13: 978-0-8160-7207-1
ISBN-10: 0-8160-7207-8
1. Medicine—History—15th century—Popular works. 2. Medicine—History—16th century—Popular works. 3. Medicine—History—17th century—Popular works. 4. Discoveries in science—History—Popular works. I. Title.

R146.K45 2010
610.9—dc22 2008055603

Text design by Annie O'Donnell
Illustrations by Bobbi McCutcheon
Photo research by Elizabeth H. Oakes

Printed in the United States of America

Bang Hermitage 10 9 8 7 6 5 4 3 2 1

This book is printed on acid-free paper.

CONTENTS

PREFACE

"You have to know the past to understand the present."
—*American scientist Carl Sagan (1934–96)*

The history of medicine offers a fascinating lens through which to view humankind. Maintaining good health, overcoming disease, and caring for wounds and broken bones was as important to primitive people as it is to us today, and every civilization participated in efforts to keep its population healthy. As scientists continue to study the past, they are finding more and more information about how early civilizations coped with health problems, and they are gaining greater understanding of how health practitioners in earlier times made their discoveries. This information contributes to our understanding today of the science of medicine and healing.

In many ways, medicine is a very young science. Until the mid-19th century, no one knew of the existence of germs, so as a result, any solutions that healers might have tried could not address the root cause of many illnesses. Yet for several thousand years, medicine has been practiced, often quite successfully. While progress in any field is never linear (very early, nothing was written down; later, it may have been written down, but there was little intra-community communication), readers will see that some civilizations made great advances in certain health-related areas only to see the knowledge forgotten or ignored after the civilization faded. Two early examples of this are Hippocrates' patient-centered healing philosophy and the amazing contributions of the Romans to public health through water-delivery and waste-removal systems. This knowledge was lost and had to be regained later.

The six volumes in the History of Medicine set are written to stand alone, but combined, the set presents the entire sweep of the history of medicine. It is written to put into perspective for high school students and the general public how and when various medical discoveries were made and how that information affected health care of the time period. The set starts with primitive humans and concludes with a final volume that presents readers with the very vital information they will need as they must answer society's questions of the future about everything from understanding one's personal risk of certain diseases to the ethics of organ transplants and the increasingly complex questions about preservation of life.

Each volume is interdisciplinary, blending discussions of the history, biology, chemistry, medicine and economic issues and public policy that are associated with each topic. *Early Civilizations,* the first volume, presents new research about very old cultures because modern technology has yielded new information on the study of ancient civilizations. The healing practices of primitive humans and of the ancient civilizations in India and China are outlined, and this volume describes the many contributions of the Greeks and Romans, including Hippocrates' patient-centric approach to illness and how the Romans improved public health.

The Middle Ages addresses the religious influence on the practice of medicine and the eventual growth of universities that provided a medical education. During the Middle Ages, sanitation became a major issue, and necessity eventually drove improvements to public health. Women also made contributions to the medical field during this time. *The Middle Ages* describes the manner in which medieval society coped with the Black Death (bubonic plague) and leprosy, as illustrative of the medical thinking of this era. The volume concludes with information on the golden age of Islamic medicine, during which considerable medical progress was made.

The Scientific Revolution and Medicine describes how disease flourished because of an increase in population, and the book describes the numerous discoveries that were an important aspect of this time. The volume explains the progress made by Andreas Vesalius (1514–64) who transformed Western concepts of the structure of the human body; William Harvey (1578–1657), who studied and wrote about the circulation of the human blood; and Ambroise Paré (1510–90), who was a leader in surgery. Syphilis was a major scourge of this time, and the way that society coped with what seemed to be a new illness is explained. Not all beliefs of this time were progressive, and the occult sciences of astrology and alchemy were an important influence in medicine, despite scientific advances.

Old World and New describes what was happening in the colonies as America was being settled and examines the illnesses that beset them and the way in which they were treated. However, before leaving the Old World, there are several important figures who will be introduced: Thomas Sydenham (1624–89) who was known as the English Hippocrates, Herman Boerhaave (1668–1738) who revitalized the teaching of clinical medicine, and Johann Peter Frank (1745–1821) who was an early proponent of the public health movement.

Medicine Becomes a Science begins during the era in which scientists discovered that bacteria was the cause of illness. Until 150 years ago, scientists had no idea why people became ill. This volume describes the evolution of "germ theory" and describes advances that followed quickly after bacteria was identified, including vaccinations, antibiotics, and an understanding of the importance of cleanliness. Evidence-based medicine is introduced as are medical discoveries from the battlefield.

Medicine Today examines the current state of medicine and reflects how DNA, genetic testing, nanotechnology, and stem cell research all hold the promise of enormous developments within the course of the next few years. It provides a framework for teachers and students to understand better the news stories that are sure to be written on these various topics: What are stem cells, and why is investigating them so important to scientists? And what is nanotechnology? Should genetic testing be permitted? Each of the issues discussed are placed in context of the ethical issues surrounding it.

Each volume within the History of Medicine set includes an index, a chronology of notable events, a glossary of significant terms and concepts, a helpful list of Internet resources, and an array of historical and current print sources for further research. Photographs, tables, and line art accompany the text.

I am a science and medical writer with the good fortune to be assigned this set. For a number of years I have written books in collaboration with physicians who wanted to share their medical knowledge with laypeople, and this has provided an excellent background in understanding the science and medicine of good health. In addition, I am a frequent guest at middle and high schools and at public libraries addressing audiences on the history of U.S. presidential election days, and this regular experience with students keeps me fresh when it comes to understanding how best to convey information to these audiences.

What is happening in the world of medicine and health technology today may affect the career choices of many, and it will affect the health care of all, so the topics are of vital importance. In addition, the public health policies under consideration (what medicines to develop, whether to permit stem cell research, what health records to put online, and how and when to use what types of technology, etc.) will have a big impact on all people in the future. These subjects are in the news daily, and students who can turn to authoritative science volumes on the topic will be better prepared to understand the story behind the news.

ACKNOWLEDGMENTS

This book as well as the others in the series was made possible because of the guidance, inspiration, and advice offered by many generous individuals who have helped me better understand science and medicine and their histories. I would like to express my heartfelt appreciation to Frank Darmstadt, whose vision and enthusiastic encouragement, patience, and support helped shape the series and saw it through to completion. Thank you, too, to the Facts On File staff members who worked on this set.

The line art and the photographs for the entire set were provided by two very helpful professionals—artist Bobbi McCutcheon provided all the line art; she frequently reached out to me from her office in Juneau, Alaska, to offer very welcome advice and support as we worked through the complexities of the renderings. A very warm thank you to Elizabeth Oakes for finding a wealth of wonderful photographs that helped bring the information to life. Carol Sailors got me off to a great start, and Carole Johnson kept me sane by providing able help on the back matter of all the books. Agent Bob Diforio has remained steadfast in his shepherding of the work.

I also want to acknowledge the wonderful archive collections that have provided information for the book. Without such places as the Sophia Smith Collection at the Smith College library, first-hand accounts of Civil War battlefield treatments or reports such as Lillian Gilbreth's on helping the disabled after World War I would be lost to history.

INTRODUCTION

[W]e shall free [medicine] from its worst errors. Not by following that which those of old taught, but by our own observation of nature, confirmed by extensive practice and long experience.

—From a pamphlet written by Paracelsus, ca. 1530

The era from 1450 to 1700 encompasses the time known as the Renaissance (from the French, *renaissance,* meaning "rebirth"), though some historians prefer to call this time "Early Modern" to dim the indication that the Renaissance was a "golden age." While there were definite societal gains from the feudalism of the Middle Ages, it was still a time filled with poverty, warfare, and oppression.

Accurately used, Renaissance describes a cultural movement that began in Italy in the late 14th century (the end of the Middle Ages) and eventually spread throughout Europe, lasting until the 18th century. The movement revived the importance of using classical learning as a base and also a stepping-stone to explore and question all types of issues. This approach was revolutionary, coming as it did after the Middle Ages where religion and superstition dominated all thinking and stalled the pursuit of new ideas.

As dissatisfaction with the prevailing religious practices began to fester, such men as Martin Luther (1483–1546) began to question the tenets of the Catholic Church. Luther and others became unfavorably impressed by the "selling" of church positions and other acts of corruption that had become a part of the era. This grew into the movement known as the Protestant Reformation and resulted in several offshoots of the Catholic Church. Because the church had been so influential in providing background for methods of healing, this shake-up in the hierarchy was to have its effect on medicine by spurring the asking of questions about medical issues. The willingness to study and explore the human body, as written about in 1543 in Vesalius's *De humani corporis fabrica* (On the fabric of the human body), is a perfect example of how medicine benefited from the new belief in the importance of asking questions. (See chapter 2.)

The questioning of everything from religious doctrines to styles of government to the understanding of the way the world works led to many significant developments, but perhaps the most important one actually concerned not a specific discovery but rather a process of discovery, the scientific method. This method was a process for experimentation that was used to explore observations and answer questions. Scientists learned that they could test cause and effect by altering variables in any subject under study, and, in doing so, they could increase their knowledge as to how something worked. This new methodology led to great developments in the fields of astronomy, physics, biology, and anatomy. Among them were the following:

- Nicolaus Copernicus (Mikolaj Kopernik) (1473–1543) advanced a *heliocentric* theory of cosmology when his *De revolutionibus orbium coelestium* (On the revolutions of the heavenly spheres) was published in 1543. The scientists of the time came to understand that the Sun, not the Earth as Aristotle had taught, was the center of the solar system.
- William Gilbert (1544–1603), an English physician who attended to both Elizabeth I and James I, laid the foundation for the theory of magnetism and electricity.
- Tycho Brahe (1546–1601), a Danish astronomer, made extensive studies and accurate observation of the planets without any magnifying device for seeing the heavens. His work laid the foundation for Johannes Kepler (1571–1630), a German astronomer who succeeded Brahe at an observatory that had been built for Brahe. Kepler did revolutionary work in the understanding of planetary

motion. He also developed a theory of light that explained vision, so he is sometimes referred to as the founder of modern optics.

- Sir Francis Bacon (1561–1626), a British philosopher and author, wrote *Novum Organum* (1620) in Latin, presenting a new systematic analysis of knowledge that was an improvement over Aristotle's method of deductive reasoning.
- Galileo Galilei (1564–1642), an Italian mathematician, astronomer, and physicist, introduced theories on gravity and motion that were later formalized by Newton. He also pioneered experiments that were then analyzed mathematically and improved a refracting telescope for astronomical use, which led to some very important astronomical discoveries.
- Scientists began to realize that Aristotle's theory that everything was made up of earth, water, air, and fire was too simple, that there was more that needed to be understood. René Descartes (1596–1650) began to theorize that the world was made up of particles of matter, a new concept for this time.
- Antoni van Leeuwenhoek (1632–1723), a Dutch cloth merchant, constructed powerful single-lens microscopes in his free time, and he made extensive observations that were published in about 1660 that opened the world of "micro" discoveries. (See chapter 5.)
- Sir Isaac Newton (1642–1727) came to realize that there were physical laws that governed motion of everything, regardless of weight, and his theories finally replaced Aristotle's concept of motion. (Aristotle had taught that heavy bodies moved straight down, light bodies moved straight up, and ethereal bodies moved in a circular motion.) Newton also believed that any scientific theory should be coupled with rigorous experimentation, which has been vital to modern science.

- William Harvey (1578–1657) provided scientists with evidence that finally overrode Galen's theory of blood circulation. (See chapter 4.)

Chapter 1 establishes the medical practices of the early 16th century and introduces Paracelsus, one of the first physicians to forcefully reject Galen. At about this same time, Leonardo da Vinci was creating unparalleled drawings of the human anatomy; yet they were not destined to be discovered and appreciated during his lifetime. Chapter 2 outlines the progress that was made in the study of human anatomy, a field that finally expands as the church begins to loosen its rules against dissections. Surgery during the Middle Ages was a high-risk type of treatment, but the use of gunpowder in battles during the 15th century necessitated that physicians begin to learn more about surgical wound-healing, and chapter 3 explains how this happened. In chapter 4, Galen's theory of blood circulation is finally debunked, and William Harvey—and some of those who followed him—put forward a concept that described accurately how blood flows through the human body. The invention of the microscope was a huge improvement in tools for medical study, but the first really good microscope was created by a cloth merchant whose discovery is explained in chapter 5. Chapter 6 examines syphilis, felt to be a new disease of the day, and by discussing the nature of both the illness and the treatment, the chapter illuminates a great deal about the attitude toward medicine of the time. Just as world explorers of this time brought back such illnesses as syphilis, they also brought back remedies. Chapter 7 alternates between what was happening in Europe and what was being discovered and brought back from the New World. Chapter 8 assesses medicine at the end of the 17th century. While great gains in knowledge had been made, scientists still had no understanding of what caused disease. As a result, bloodletting, astrological predictions, and alchemy—in combination with some of the medical improvements that had come about—were still the order of the day.

The Scientific Revolution and Medicine: 1450–1700 illuminates what occurred during the Scientific Revolution that affected future developments in medicine. The back matter contains a chronology, a glossary, and an array of historical and current sources for further research. These sections should prove especially helpful for readers who need additional information on specific terms, topics, and developments in medical science.

This book is a vital addition to the literature on the Scientific Revolution because it puts into perspective the medical discoveries of the period and provides readers with a better understanding of the accomplishments of the time. While physicians of this era did not yet know the cause of disease, they had begun to make many advances that were to be key to medical improvements to come.

1

Medicine: Ready for a New Start

Most historians date the beginning of the Scientific Revolution to 1543, the date when Nicolaus Copernicus (Mikolaj Kopernik) published *De revolutionibus orbium coelestium* (On the revolution of the heavenly spheres) and Andreas Vesalius published *De humani corporis fabrica* (On the fabric of the human body). These two men and their works were part of a major transformation in scientific ideas in many fields, including physics, *astronomy,* and biology. As a result of all these changes in so many areas, the groundwork was laid for the development of what is now considered modern science.

As with any type of transition, a great deal of societal shifting has to take place to prepare for a major transformation, and while it is virtually impossible to identify a specific event that started the cascade of change, certainly the expansion of the known world was an early factor. Shipbuilders began to develop vessels that permitted longer and more ambitious sea travel, so sailors began to return with fantastic tales of what they saw and to bring back souvenirs of their adventures. This awakened a new interest in learning, which encouraged education. While the number of university-educated men remained quite small, their very existence provided a new elite willing to examine issues differently. The rise in university training in medicine brought about a renewed interest in Greek medical thought, and the documents preserved by Islamic scholars were being translated into Latin to provide scholarly background.

The atmosphere of change in so many aspects of society—from explorers traveling back with reports of never-before-seen lands to economic and religious upheaval—created an environment that led to questioning the past. Even the church became subject to criticism as such people as Martin Luther began to point out the abuses of power that the church permitted its leaders.

In addition, there was a health-related factor that turned Europe upside down. The Black Death, which shrouded the Continent in 1347–48, was one of the deadliest pandemics in human history, wiping out from 30 to as high as 60 percent of a town's population. As a result of this high rate of fatality, European society had to reorganize economically. As more of the lower class people fell ill, feudal lords no longer held the upper hand as they had fewer people available to do their bidding; tenant farmers began to ask for ownership, which brought about an eventual shift in economic distribution. This led to significant changes in societal structure.

The Black Death also brought about new thinking on the issue of autopsies, which had long been forbidden by the church and as a result held back medical progress because of the inability for *physicians* to study *anatomy.* Religious reverence for the human body had always held that it was a *sacrilege* to cut into the body for the purpose of study, and doctors faced legal action and public censure if they attempted to perform autopsies. As towns were wiped out by the Black Death and bodies were left to pile up in the streets because no one had the time to bury them, religious

---

(Opposite) At the beginning of the early modern world, civilizations were very isolated, and trips from Europe to the various populated areas took months, sometimes years.

The Early Modern World (1500s–1600s)

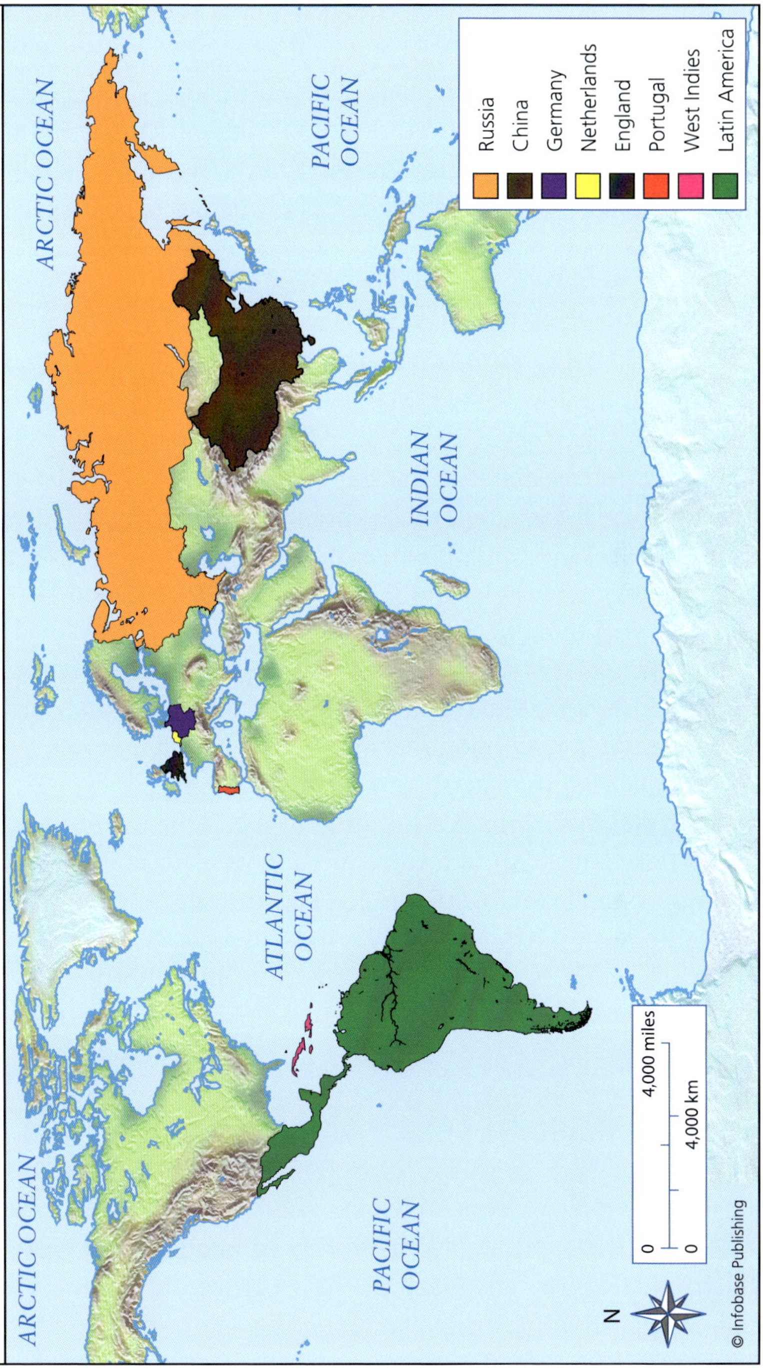

leaders wanted to know what was causing this terrible disease. As a result, they began to permit postmortem examinations of plague victims. It took another 200 years before autopsies were conducted more regularly, and in 1537 Pope Clement VII finally permitted human *dissections* in anatomy classes. Had the plague been less severe, perhaps this change in attitude would have taken even longer.

In the 21st-century era of specialization, one particular aspect of these leaders should be noted. The artists and leaders who contributed their inventions, thoughts, and writings were notably versatile and multifaceted. Many were interested in both science and art, and they made major contributions in more than one area. World-renowned artist Leonardo da Vinci is today remembered primarily for his art, but his notebooks reveal brilliance in several fields. Among his accomplishments were an accurate description of the science behind plate tectonics (at a time when the peasant class still thought the world was flat), and he developed ideas for amazing inventions such as a hydraulic lift. This chapter will highlight his contributions to anatomical drawings, and, although these were not even known about during his lifetime, they are so remarkable that they merit attention even today.

This chapter examines the state of medicine in the early part of the 16th century, and it introduces Paracelsus, a major force in moving beyond Galen's theories. Leonardo da Vinci's studies on the anatomy of the human body will be examined, and the notable influence on medicine of the invention of the printing press will be highlighted.

GALENIC MEDICINE STILL PREVAILS

In the early 16th century, physicians still relied on the medical ideas of the Greek physician Galen (129–199 C.E.), whose theories about medicine still guided all forms of analysis and treatment. Galen made many advances in the work he did during his lifetime, and, had his theories been "stepping-stones" to other things, he would have been forever remembered for his great advances

The Four Humors

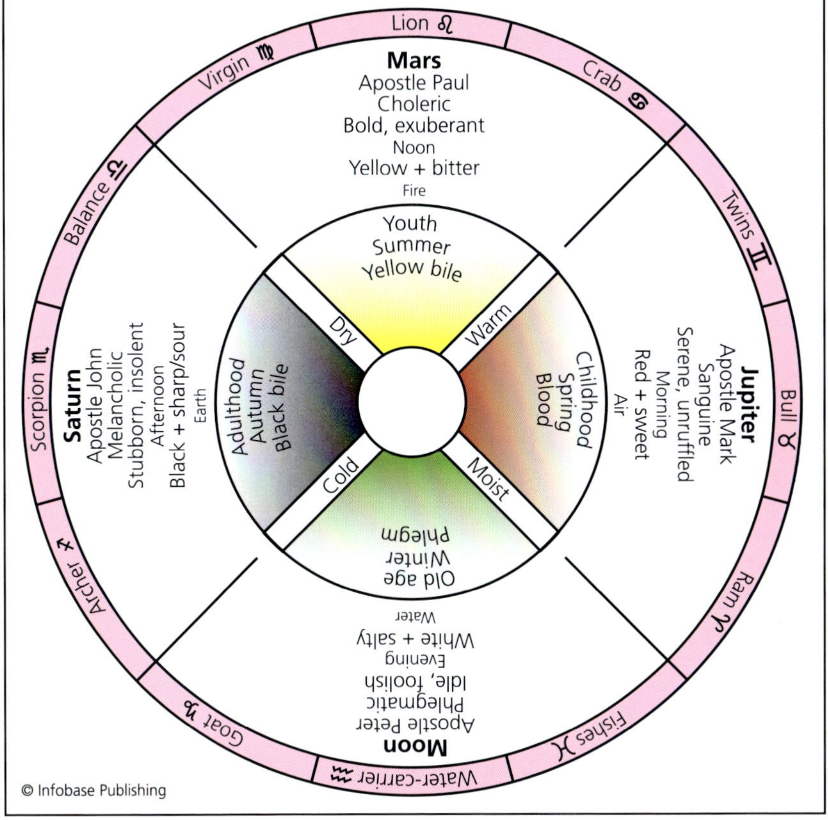

The medical community continued to believe in the value of balancing the four humors.

in medicine. Unfortunately, Galen collected a huge following of believers, and his bombastic approach to anyone who questioned him made others view his theories as unassailable. As a result, Galen's methodologies prevailed over an amazing 1,500-year time span.

The importance of balancing the four humors (blood, phlegm, black bile, and yellow bile) was one of Galen's notions that prevailed. Galen recommended specific diets to help maintain humoral balance, and purging and bloodletting were important solutions if someone fell ill. Galen was fascinated by anatomy, and he dissected daily, but because human dissection was forbidden during his time, he performed his work on various animals whose anatomy he believed was similar to the human body. Unfortunately, his writings did not reflect the nature of the subject he was dissecting, so those who followed him were misguided by a good portion of the information Galen noted about anatomy.

Galen made good progress in the study of the blood, though there were still misconceptions. He realized that the arteries carried blood, not air (pneuma) as was commonly believed, and he came to understand the importance of the pulse in assessing a person's state of health. Galen, however, argued that blood was continuously made by the liver and was used up. This validated the use of bloodletting. If blood was created continually, then there was no problem with draining it in measured amounts.

Galen maintained his own garden to create medicines. He created both plant- and animal-based medicines, and many of his concoctions consisted of an overwhelming number of ingredients. Galen's "theriac" was the best known, and Galen wrote an entire book about making it and what it could be used for. It was made of at least 64 ingredients including flesh from a viper. Theriac, as well as many of Galen's other mixtures, continued to be used medicinally as late as the 19th century.

During his day, Galen did an amazing amount of work to move medical knowledge forward. Western society's misfortune was that few could overcome the power of the Galenic beliefs. Nearly 1,500 years later, physicians were still locked into health theories that were rarely helpful and sometimes harmful. In addition, because the ideas were staunchly supported, there was little movement to experiment and learn anything new.

TWO OTHER PRACTICES OF THE DAY

Medically speaking, this was a time when magic still overpowered rationalism, and there were two other areas that fascinated physicians. The first was medical treatment based on *astrology*, and

Medical Care and Astrology

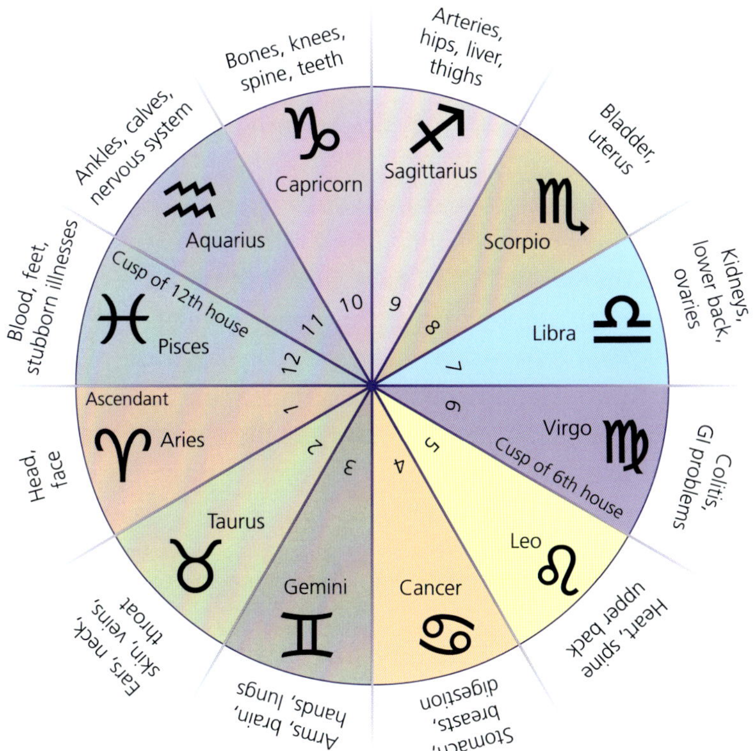

Physicians believed certain astrological signs governed specific parts of the body, and they also took into account a patient's astrological sign before determining a treatment.

the second was the practice of *alchemy*. Both of these areas were very influential. While doctors no longer treat based on a patient's astrological sign or the star configuration when they became ill, many people today still follow their horoscopes and give passing credence to the thought that their lives may be influenced by the hour at which they were born. While alchemy was largely a misguided idea of turning one substance—usually a metal—into something completely different, it spurred on the idea of mixing things up, and, in the process, more and more men began to pursue what is now called chemistry.

Astrological medicine was guided by a very complex set of rules, and it was based on the assumption that the motion of the heavenly bodies influenced human health. Using astrological medicine in patient care began with the physician trying to ascertain the exact moment that a person became ill. The next step involved studying the heavens to predict what the course of the illness would be. The Sun was thought to rule chronic diseases, and melancholy was blamed on Saturn. The Moon governed the flow of blood, so the position of the Moon dictated the proper time and method for bloodletting and any other type of surgery. Charms were often used as part of the healing process. Because this type of medicine was without merit, patients were rarely helped unless they were going to pull through anyway. Over time, a growing number of physicians began to turn away from and openly condemn astrological medicine.

Alchemy is generally known as a method to transform base metals into gold, but at that time alchemy was broader than that. The Chinese viewed it as a way to change certain ingredients into *elixirs* to provide good health, and in the West during the High Middle Ages, alchemy was adapted as a method for preparing medicines. Some 16th-century scientists held alchemists in high esteem, feeling that alchemists were pioneers of chemistry; others thought that they were charlatans.

PARACELSUS LEADS THE WAY

To begin to move away from medicine of the past takes someone brave who does not particularly worry about currying favor with others, and in the early part of the 16th century, Europe had that type of *iconoclast* in the form of Paracelsus, who was born as Phillip von Hohenheim (1493–1541). He was a brilliant but controversial figure in the world of medicine and introduced fascinating new theories that became very influential. His ideas were slow to take hold because he was arrogant and not well liked by other physicians.

Paracelsus was born in the Tirol mining district of what is now Austria, and he is thought to have gained a medical degree at the University of Ferrara where he became enamored of the teachings of Hippocrates. He took the name Philippus Aureolus Theophrastus Bombastus von Hohenheim, signaling that either he or his father had grandiose visions of what he was to accomplish. Aureolus was the name of a famed alchemist, and Theophrastus was Aristotle's successor, a great philosopher, and the first systematic botanist. When he shortened his long name to Paracelsus, it meant "greater than Celsus." (Aulus Cornelius Celsus was one of the great encyclopedists of the first century C.E.)

Other physicians of the day were beginning to study anatomy, but Paracelsus felt one could learn nothing from the dead. He was convinced that the only way to learn about illness was by studying the living body. He also valued what he could learn from healers, and between 1510 and 1524, he traveled throughout Europe, Russia, and the Middle East, where he absorbed the information shared with him by barber-surgeons, midwives, and folk healers. Eventually, he acquired a background in medical science and chemistry of the time, and he also learned about the occult, astrology, and alchemy. Paracelsus was frequently seen in the alchemist's leather apron rather than academic robes. He loved experimenting with chemistry, and he turned it into a performance art and dazzled audiences with his chemical wizardry.

A constant learner, Paracelsus realized that there was no better opportunity to observe the human body under stress than on the battlefield. He had learned enough surgery that he felt qualified to follow the Habsburg armies that were fighting in Italy and Scandinavia to provide care. As he helped manage the soldiers' wounds, he began to understand that infection was often the ultimate villain in taking the lives of the wounded young men. During this time, the treatment of choice for injuries sustained in battle often involved covering the wounds with boiling oil, dung, and other substances. Infection was often the result. Paracelsus saw the senselessness of what was being done, so he came up with a substitute theory that he hoped would divert the surgeons. He suggested that the concocted mixture should go on the weapon that caused the wound, and, in so doing, this treatment would be curative. (Healing through magic was still an active belief, so this would not have seemed as far-fetched as it might seem today.) Paracelsus's theory proved helpful. The soldiers' wounds were cleaned and then left to self-heal. Because the mixtures used were so inappropriate for wound care, this method was far preferable to putting these misunderstood agents directly onto the wounds.

Paracelsus's status became exalted in the early 16th century when he was asked to treat humanist publisher Johannes Froben, who had a bad infection of his right leg. Paracelsus crafted a comprehensive plan of treatment, and Froben lived. In gratitude, the city council of Basel, Switzerland, made him an official physician of the city, and he was encouraged to write, teach, and experiment.

Paracelsus, a most controversial figure in medical history, is shown in one of his many "chemical kitchens," about to embark upon one of his mystical and frequently vitriolic writings. His laboratory, desk, and manuscript piles reflect his habitual disorderliness. Alchemical experimentation, mystical speculation, prolific writing, and empirical practice of medicine were equally confused facets of his life. *(Department of Library Sciences, Christian Medical College—Vellore, History of Medicine Picture Collection)*

Eight months later, he was told that he was no longer welcome to stay. Historians cite two possible reasons for his banishment: Students at the university had created a bonfire in celebration of a religious holiday, and Paracelsus threw in the *Canon of Avicenna (Ibn Sina)* as an expression of his disdain for the work. To other physicians, this was a sacrilege. The other possibility had to do with Paracelsus's manifesto that essentially declared war on medicine. He claimed that doctors' prescriptions were, at best, misguided and useless, and more likely were contaminated and dangerous. He capped that off with the ultimate insult to the profession: He noted that physicians' services were overpriced.

NEW DISCOVERIES CHALLENGE OLD IDEAS

Paracelsus was the first to step away definitively from Galen's theories, and in the process, he made the following significant contributions to medicine:

1. He followed Hippocrates' observation-based medicine, believing that each disease was a separate entity that resulted from agents outside the body that could be cured with a treatment that addressed those symptoms. (This was a good first step on the way to germ theory.) His beliefs also caused him to reject Galen's humoral balance theory, a belief that had dominated for the past 1,500 years.
2. His study of alchemy under Islamic chemists led him away from plant-based mixtures that were popular at the time, and Paracelsus introduced the idea that medicines could be mixed from other compounds. He used the principles of alchemy—the extraction of pure metals from ores, the production and use of powerful solvents, evaporation, precipitation, and distillation—to make medications. In combination with plant extracts, he mixed arsenic, lead, sulphur, copper, sulphate, zinc, mercury, and antimony. He knew that these metals could also be poisonous, and he noted

that the secret was in the dosage. This work paved the way for a more serious application of chemistry to medicine.

3. His work as a military surgeon gave him great respect for surgery as an art, and he fought against the idea that surgery was an inferior branch of medicine. He wrote *Die grosse Wundartzney* (Great surgery book) that was published in 1536.
4. Paracelsus, who was raised in a mining community and observed his father treating the workers, came to realize that smelters, miners, and metallurgists all had certain illnesses because their lungs and skin absorbed noxious pollutants. He eventually wrote a book on miners' disease and recognized that it was a metabolic disease.
5. In 1522, Paracelsus is thought to have learned a peasant remedy to prevent *smallpox*. Paracelsus visited Constantinople where peasant women were using a method of inoculation a full two centuries before Lady Montagu (1689–1762), who introduced it to England after learning of it while her husband was ambassador to Turkey. This was also way before the English physician Edward Jenner (1749–1823) formalized the process. Paracelsus learned about pulverizing the scabs of smallpox lesions for people to inhale. He tried it with other diseases, but success in vaccinating against other illnesses did not prove successful at that time.
6. He was also the first to manage effectively the congenital form of syphilis. In Nürnberg (Nuremberg), he was asked to demonstrate his theories by curing syphilis when sailors from Columbus's voyage came home with it. He cured nine out of 14 cases using mercury. He wrote about the illness and the remedy, and mercury remained the treatment of choice until 1909 when Paul Ehrlich discovered Salversan, an arsenic compound.
7. Paracelsus believed in nature's healing methods and noted that "If you prevent infection, nature will heal the wound all by herself."

8. He believed that doctors should treat rich and poor alike, and that a graded fee system, with the poor being treated for free while the wealthy paid more, evened out the earnings of doctors.

Paracelsus died at a young age. There is speculation that other physicians had him attacked, leading to the fall from which he died.

The work of Paracelsus highlights the divide between the old theories supporting the universe and the new ideas that appealed to patients as well as those physicians who were prepared to challenge the old ideas. Because Paracelsus was a controversial character who knew little about the art of explaining and nothing at all about persuasion, his theories had a very bumpy path, but eventually they were picked up by others who could more smoothly convey Paracelsus's wisdom. Nonetheless, the Scientific Revolution had begun, leading to reevaluations in many areas.

LEONARDO DA VINCI (1452–1519): CONTRIBUTIONS TO MEDICAL KNOWLEDGE

Leonardo da Vinci is best remembered today for his paintings. Though there are only 17 known works—not all of them completed—some of his paintings, the *Mona Lisa* and *The Last Supper* among them, are the most famous in the world. His drawing of *Vitruvian Man,* described later in this chapter, is *iconic.*

Contemporaries knew that he was a highly gifted individual who contributed to many fields, including architecture, technology, military weaponry and fortifications, human aviation, and botany, and he developed a basic explanation of plate tectonics. All of these ideas were well ahead of their time. Less well understood—and basically unknown during his lifetime—were his contributions to the field of medicine. Unbelievably beautiful and anatomically accurate drawings of various parts of the human body filled many of Leonardo's notebooks, but this work was not discovered by others until after his death. As a result, his incredible step forward in the field of anatomy remained unknown until at least the 1650s.

Unfinished painting of St. Jerome in the wilderness by da Vinci, ca. 1480 *(The Yorck Project)*

Leonardo's Life

Leonardo was the illegitimate son of a Florentine notary, Piero da Vinci. He was born in the Vinci region of Florence, so he would have been known as Leonardo di ser Piero da Vinci. When he was 14, Leonardo was apprenticed to one of the most successful artists of the day, Andrea di Cione, known as Verrocchio. Verrocchio believed strongly that his apprentices needed to master a wide range of technical skills as well as to undertake serious study of drawing, painting, and sculpting. Verrocchio emphasized that his pupils study anatomy, and Leonardo showed an immediate gift for topographic anatomy, drawing many studies of muscles, *tendons,* and other visible features.

Though his only formal education was in art, Leonardo was fascinated by a wide range of subjects and taught himself in fields as diverse as mathematics and Latin. The Renaissance was a time when science and art were not considered polar opposites. The notebooks that contained his work were filled with thousands of pages of notes and sketches on many subjects, ranging from studies of the inventions that he was conceptualizing (including a helicopter and various forms of hydraulic lifts), and his anatomical studies, which were significant to the world of medicine. His drawings of the human anatomy are unrivaled.

His Interest in Anatomy

During this era, the Roman Catholic Church forbade human dissection, believing that it violated the sanctity of the human body. However, when a Veronese anatomist, Marcantonio della Torre, gained special permission to perform dissections, he asked Leonardo to work alongside him to prepare illustrations for a text on anatomy. When Della Torre died unexpectedly, Leonardo assumed both tasks, performing the dissections and then working on the illustrations. Because he was not the one who had gained permission, he worked in secrecy in the cathedral cellar of the mortuary of Santo Sprito in Florence, dissecting and drawing as many as 30 human bodies.

Leonardo drew many studies of the human skeleton and its parts, as well as muscles and *sinews,* the heart and vascular network, the reproductive system, and other internal organs. He

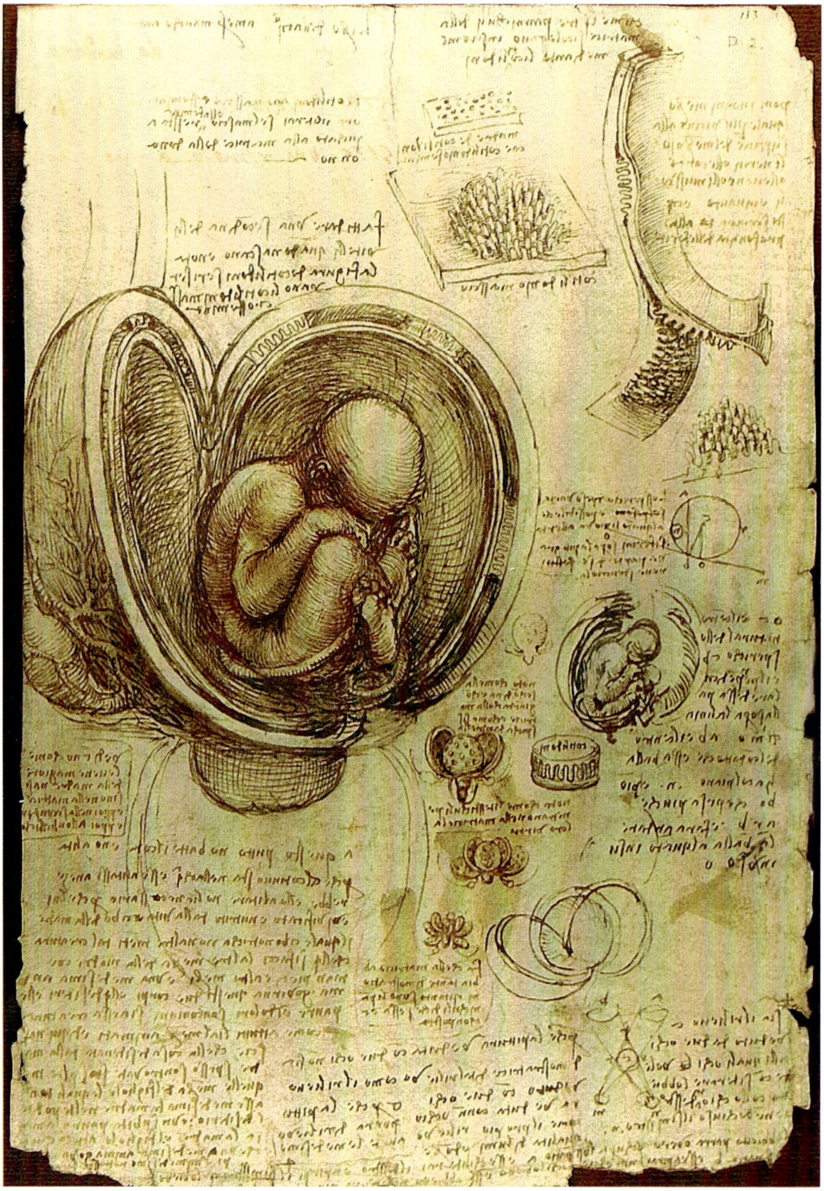

Da Vinci Studies of Embryos, ca. 1510 *(Luc Viatour)*

made one of the first scientific drawings of a fetus *in utero.* While the topographical studies were notable, Leonardo's dedication to observing and recording individual parts of the body as they performed mechanical activity was the feature that made his work so exceptional. He probed the brain, the heart, and the lungs, and he found ways to draw transparent layers to depict the internal organs and how they functioned. He also observed and recorded the effects of age, emotion, and disease on *physiology.*

His anatomical studies of animals permitted additional study, and he worked out ways to expand his knowledge. He injected hot wax into the brain of an ox, which provided him with a model of the ventricles. This represented the first known use of a solidifying medium to define the shape and size of an internal body structure. He developed an original mechanistic model of sensory physiology and worked at researching how the brain processed visual and other sensory input.

He seemed to read widely, and his interest in dissection may have been inspired by reading Galen. He differed from Galen, however, in understanding that human dissection was vital to understanding human anatomy. (Galen felt other living creatures could be studied instead.) Though Leonardo differed from Galen on many issues, he maintained the description of the circulatory system that Galen provided, indicating that "pores" between the ventricles permitted the blood to travel between the two sections of the heart. Leonardo's illustrations do not reflect these pores between the ventricles, but Galen was so revered that even when the anatomy did not fit with the theory, Galen was held to be correct.

Many of Leonardo's drawings were done on various-sized loose pieces of paper, and it is thought that they were collected into notebooks by one of his students. Though the material appeared to be intended for publication, it is not clear why that never occurred. Leonardo was known to be a procrastinator so it may have been that he never got around to it, or it could have been that his lack of a formal education in anything but art—and hence his lack of formal education in mathematics and Latin—left him feeling that he did not have the right credentials to publish in a more scientific field.

His inventions and anatomical drawings were usually accompanied by Leonardo's explanations of what he was drawing. These notations were written in mirror-image cursive. It was originally thought that Leonardo intended the notations to be somewhat secretively written, but later it was noted that Leonardo wrote with his left hand, and so it was probably simply a practical solution to prevent smearing. It would have been far easier to write from right to left with a nib pen if he were using his left hand.

In 1651 (almost 150 years after his death), many of his anatomical drawings were published for the first time as part of a treatise on painting. The wealth of Leonardo's anatomical studies that have survived forged the basic principles of modern scientific illustration.

AN UNDERSTANDING OF PROPORTIONS

Though Leonardo's anatomical studies were kept private, he published some of his observations of human proportions, most notably *Vitruvian Man.* This work was quite fascinating because it so perfectly captured the proportions of the human body.

Proportions of the Human Figure Understood

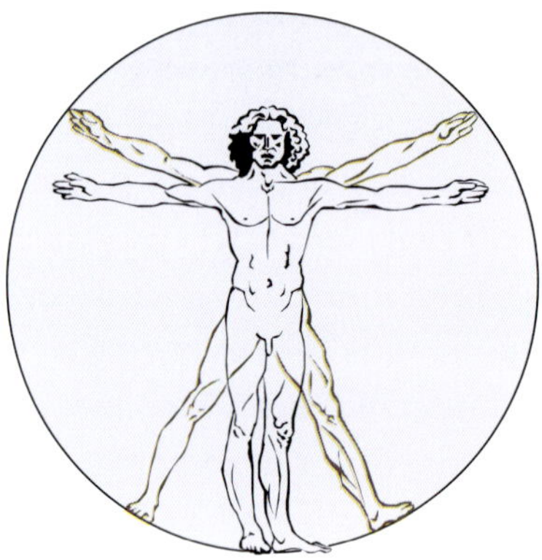

Leonardo da Vinci was the first to understand the proportions of the human body.

Leonardo took the proportional theories of Vitruvius, the first century B.C.E. Roman architect, and imposed the principles of geometry on the configuration of the human body. Leonardo demonstrated that the ideal proportion of the human figure corresponds with the forms of the circle and the square. Leonardo's illustration of this theory shows that when a man places his feet firmly on the ground and stretches out his arms, he can be contained within the four

HOW THE INVENTION OF THE PRINTING PRESS CONTRIBUTED TO MEDICINE

As the medieval period drew to a close, documents in the West had to be hand-copied by scribes. The Eastern world—ancient China and later Korea—had been using more advanced printing methods involving woodblock as well as movable type printing techniques, but these had not yet filtered West.

Then in 1439, German goldsmith Johannes Gutenberg devised a method of printing using metal molds and alloys to create movable type. He found a way to use the movable type with a special press and oil-based inks, and in the process he was able to mass-produce books. For the first time, multiple copies of printed material could be created, and each one would be the same as the one before it. (Copying documents by hand was not only time-consuming but also prone to errors as mistakes were made during the copying.)

Gutenberg's invention of the printing press was to have a massive effect on society because, for the first time, information could be spread much more easily to an increasing number of people. While at first printing did not totally dominate the written word and handwritten manuscripts continued to be produced, the invention of the printing press led to the establishment of a community of scientists who could spread the word about what they were doing. Scholarly journals and books now provided accurate descriptions that could be duplicated and communicated to much wider audiences.

The printing press also brought about another significant change. As more people could have access to information, a demand grew for more material to be created in the vernacular. No longer was Latin considered the best choice for writing about medicine.

(continues)

(continued)

Three of the medical specialists who were particularly influential because they were available in print were Andreas Vesalius (1514–64), who wrote one of the most influential books on human anatomy; anatomist William Harvey (1578–1657), who was able to accurately discern how the circulatory system worked; and Hermann Boerhaave (1668–1738), who is sometimes referred to as the father of physiology. He wrote encyclopedic medical books, such as *Institutiones medicae,* that were translated into many languages.

lines of a square, but when the body was in a spread-eagle position, it could be inscribed in a circle.

CONCLUSION

As European society underwent changes in economy and religious beliefs, the groundwork was laid for new examinations of many fields, including medicine. The devastation of the Black Death led to the beginning of church-sanctioned autopsies, which greatly increased the knowledge of human anatomy. Leonardo da Vinci's contribution to anatomical knowledge was vast but not known until after his lifetime. The physician and alchemist Paracelsus did a great deal to break the restraining bonds of Galenic belief, and, as new scientists entered the field, they were able to move forward with fewer restrictions than those who had preceded them.

2

Amazing Advances in Anatomy

Beginning in the 16th century, the study of anatomy became an important foundation for Western medicine. As noted previously, the dire number of fatalities from the Black Death in the 14th century began to set the tone for a change in attitude about dissections. Initially, the church permitted autopsies to be done on plague victims solely to try to assess the cause of death, but later strictures against autopsies began to loosen. After the laws changed in 1537 and autopsies were permitted on an as-needed basis, the physicians of the day were able to study the human anatomy more regularly.

Eventually, the study of anatomy became a part of the medical school curriculum, but even then it was still difficult to obtain cadavers to dissect. The church regulated the numbers of bodies that could be made available, and since there was no refrigeration it was difficult to study a body thoroughly before it began to decay. (Even when the dissection was done within three days—fast for that time—the stench became unpleasant for both students and teachers.)

This chapter will introduce the scientists and the physicians who worked to better understand the human body. Andreas

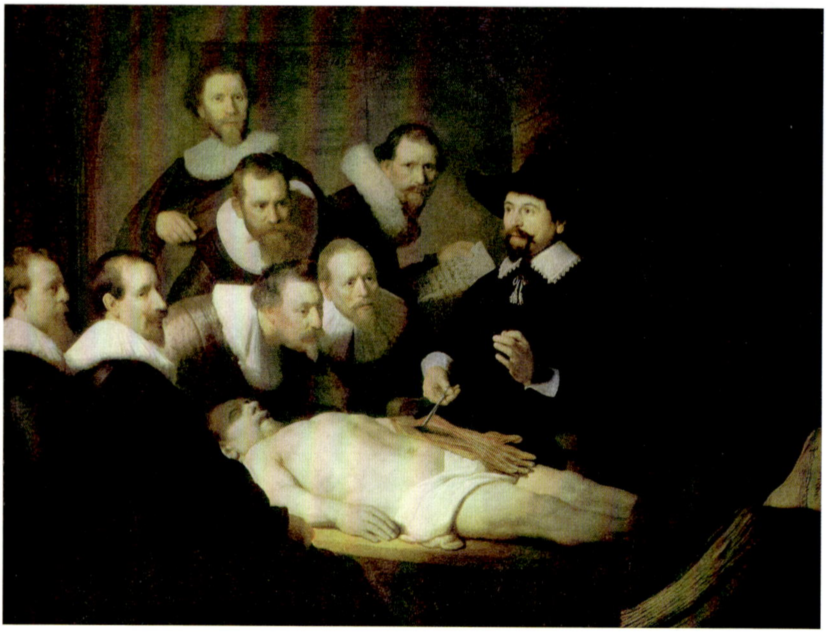

The Anatomy Lesson of Dr. Nicolaes Tulp by Rembrandt, 1632 *(The Yorck Project)*

Vesalius was the first to see that Galen's understanding of anatomy was in large measure wrong, and he was joined by several others who helped clarify the understanding of anatomy. Miguel Serveto, a theologist and physician, correctly explained pulmonary circulation, but his work was never widely acknowledged. Realdo Colombo drew needed attention to pulmonary circulation. Gabriele Falloppio (Falopius), one of Vesalius's students, succeeded him as a professor of anatomy at Padua, where he continued to explore the body's structure and made notable advances in the study of the skull, the ear, and the female genitalia. Vesalius also inspired others to more closely study the organs and how the body worked. Another who did so was Bartolomeo Eustachio (1520–74), who discovered the eustachian tube, the *suprarenals,* the thoracic duct, and the *abducens nerve.* Also, Santorio Santorio helped bring about an understanding of *metabolism.*

VESALIUS AND WHAT HE LEARNED ABOUT THE STRUCTURE OF THE HUMAN BODY

Andreas Vesalius (1514–64) was born into a family of physicians in Brussels, Belgium, and he took an early interest in how living things worked. While still a boy, he was said to have done dissections on small animals on his mother's kitchen table, which may have helped prepare him for a world where dissections were finally becoming an accepted part of medical studies.

His medical education began at the University of Louvain, followed by a move to the University of Paris in 1533 where he studied under the well-respected teacher Jacob Sylvius (1478–1555). Sylvius used dissection to study Galen, but, like his contemporaries, he saw only what Galen wanted him to see, ignoring the discrepancies between Galen's conclusions and the actual dissections. Vesalius noted the differences, and he began to speak openly about his disagreements with Galen's theories and those who taught them unquestioningly. According to the historian Lois N. Magner, author of *A History of Medicine,* Vesalius was said to have told students that they "could learn more at a butcher shop" than at a lecture by a particular professor, meaning Sylvius. Vesalius's disdain for Galen greatly angered Sylvius and other members of the faculty.

Vesalius eventually moved on to the University of Padua to complete his studies (he received a degree in December 1537) and was offered a professorship there. Vesalius continued to perform more and more animal and human dissections, and he began to notice that some of Galen's notes were true for apes and monkeys but that human skeletons did not have the same features. Galen wrote of locating a "small projection of bone upon one vertebrae of its spine." Vesalius found the additional bone mass on an ape's skeleton but could not find it on a human. He realized that Galen must have been dissecting monkeys and assumed that what he found on an ape or a monkey would hold true for humans, too. Over time, Vesalius began a full-scale assault on Galen. Vesalius arranged to conduct a side-by-side comparison for the public in Padua, dissecting an ape on one table and a human on the other. (There was no shortage of audiences for this type of thing.) He pointed out more than 200 differences between the two skeletons. The "small projection" on the vertebrae described by Galen was found only on the ape. As Vesalius had promised, the human skeleton had none.

After a brief stint in the military, Vesalius took a teaching position at the University of Venice. He ran afoul of this faculty, too, by breaking with traditional teaching methods. At this time, medical classes employed three instructors. The professor was a physician who taught the class from a raised platform, a barber-surgeon was there to perform the dissection, and an "ostensor" (meaning one who shows; from medieval Latin, *ostendere,* "to show") was there to point out the parts of the body. Vesalius preferred to fulfill all three roles, performing the dissection himself while also lecturing and pointing out what he was discussing.

Vesalius's lectures aroused high interest, and to investigate in more depth he began to take longer to perform dissections, which gave him time to investigate organs and musculature that normally had been rushed through. His work came to the attention of a judge in the Padua court system, and the judge began to award the bodies of executed criminals to Vesalius. Winter was the best time to study bodies as the cold weather slowed the pace of decay, so the judge established more executions during the colder weather, and he spread out the timing of them so that the gifted anatomist would have a steady flow of bodies to study.

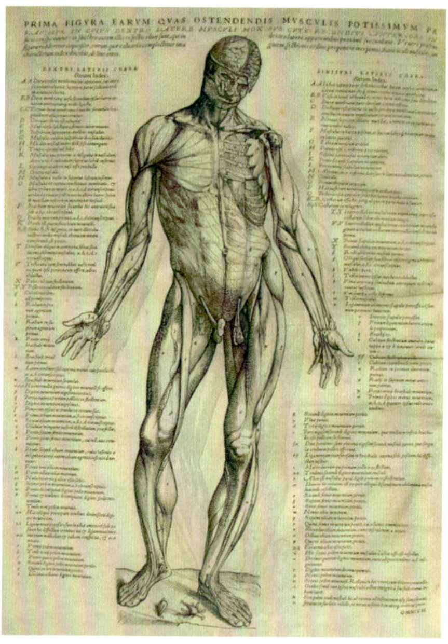

Folio 8r showing the first and second layers of muscles from the *Epitome* of Vesalius, Basel, 1543 *(University of Glasgow Library)*

In 1543, Vesalius published *De humani corporis fabrica* in an effort to inform a wider audience of his findings. At the time, this was the most accurate book on human anatomy, and it is still highly respected for both its beauty and its high level of accuracy. Further discussion of this book can be found in the following sidebar.

There Were Still Errors

Vesalius's dissections gave him an excellent understanding of anatomy, but there were still many mysteries about how the body worked, and Vesalius—like others of his day—relied on Galen's theories about blood flow, which were later found to be inaccurate. Though he did not solve the problem of how the blood traveled through the heart, he did raise the issue that the denseness of the *septum* led to the conclusion that this would have been a very unlikely process. The author Allen G. Dubus quotes Vesalius in *Man and Nature in the Renaissance:* "Not long ago I would not have dared to turn aside even a hair's breadth from Galen. But it seems to me that the septum of the heart is as thick, dense, and compact as the rest of the heart. I do not see, therefore, how even the smallest particle can be transferred from the left ventricle through the septum." (It was another 100 years before William Harvey in 1615 was able to come up with a better understanding of the movement of blood since Europeans were not aware of progress in the Islamic world.)

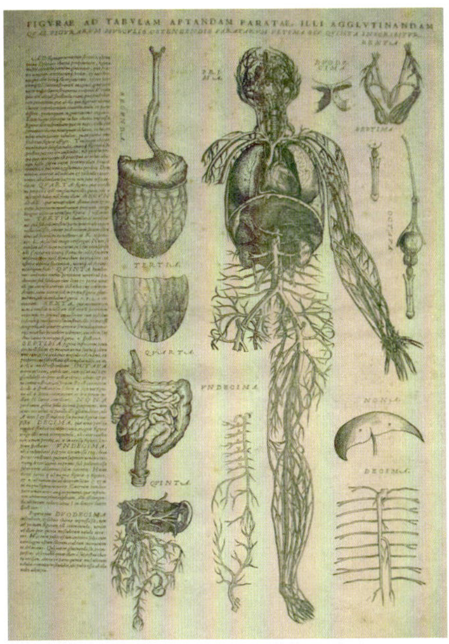

Folio 12v showing cardiovascular system and female genitalia from the *Epitome* of Vesalius, Basel, 1543 *(University of Glasgow Library)*

DE HUMANI CORPORIS FABRICA LIBRI SEPTUM

De humani corporis fabrica libri septum (On the fabric of the human body in seven books) was written by Andreas Vesalius in 1543. The writings were based on his lectures at the University of Padua. In these lectures, Vesalius broke new ground because he dissected the corpses himself, explaining what he saw along the way.

Fabrica corrected some of Galen's worst errors, including the belief that the blood originated in the liver, but Vesalius did not fully understand the circulation of the blood, so he continued to hold Galen's belief that two types of blood flowed through the body—one kind traveled the arteries; the other the veins.

Vesalius took great care with his work and selected a superior illustrator, Jan Stephen van Calcar (1499–1546) who had studied under Titian (ca. 1485–1576), a leading painter of the Italian Renaissance. Van Calcar's exactness of musculature and his depiction of organs are remarkable even by today's standards. His book provided exact descriptive illustrations of the skeleton, the muscles, the nervous system, the viscera, and the blood vessels.

Vesalius understood the benefits of his material—both the texts and the illustrations—being carefully reproduced, and he realized the benefits of having his materials copied by a

Vesalius also explored to try to identify the five-lobed liver, the seven-segmented sternum, and the horned uterus, which previous physicians had written about. Through his dissections, Vesalius demonstrated that these accounts were not accurate. In a subsequent edition of *Fabrica* that was published in 1555, Vesalius printing press rather than being copied by hand, which was time-consuming and subject to errors. He sought out the best of the Renaissance printers, Johannes Oporinus, who was well known for his meticulous work. Vesalius went to Basel, Switzerland, where Oporinus worked, so that he could carefully supervise the printing.

The success of the book provided Vesalius with money and fame. When he became physician to the Holy Roman Emperor Charles V, he dedicated the book to the ruler and presented him with the first published copy, which was bound in purple silk and contained hand-painted illustrations that only existed in this copy.

A copy of *Fabrica* that is bound in human skin was a gift to Brown University's John Hay Library by an alumnus. The cover is described as "polished to a smooth golden brown" (*Boston Globe* January 7, 2006), looking and feeling much like any leather. Binding in human skin was not uncommon in centuries past. The skin was generally obtained from criminals who were executed, from people who died in poorhouses with "no next of kin," or from medical schools where bodies were donated for study. The books that were so bound were often medical books, and the choice of binding was generally meant to honor those who furthered medical research.

returned to Galen's theory about blood flow, examining how blood traveled through pores in the septum of the heart. Vesalius also believed that the purpose of breathing was to cool the blood and that the digestive process involved some way of "cooking" the food to digest it.

Affected by Disdain

Vesalius was highly criticized for differing with Galen, and in his book *A Short History of Medicine* (1955, revised in 1982), Erwin H. Ackerknecht notes that Vesalius became frustrated by the vociferous criticism of his work. He accepted a position as court physician to Charles V, who was Holy Roman Emperor and, as Charles I, king of Spain. His responsibilities were quite demanding. Charles was not particularly well, suffering from both *gout* and asthma, and so care of the king took time. In addition, it was general practice that court physicians were also loaned out to noble families or royalty from friendly countries.

Vesalius asked permission to make a pilgrimage to the Holy Land, and it was reported that when he returned, he hoped to return to teaching. As it happened, he died before returning from the pilgrimage.

SERVETO RECOGNIZES PULMONARY CIRCULATION

Miguel Serveto (1511–53), known as Michael Servetus, was a Spanish theologian and physician who lectured and wrote on geography and astronomy, but his deepest commitment was to theology. Serveto was the first to develop a coherent understanding of pulmonary circulation. The Islamic physician Ibn an-Nafis (1213–88) had written about pulmonary circulation 300 years earlier, but most Islamic medical and scientific discoveries were unknown in Europe at this time. Though Serveto was the first of the European physicians to recognize how the system worked, he did not have the reputation or the stature that permitted him to have an impact on the medical knowledge of his day.

Religion was Serveto's prime interest, and at age 15 he entered the service of a Franciscan friar before studying medicine at the University of Paris. Though he began to practice medicine, he primarily traveled in religious circles, and this exposure made him aware of religious *dogmatism* and intolerance, and he became distressed by papal ostentation. He began to fight against these issues, but Serveto was a difficult fellow who had trouble expressing his beliefs in such a way that people could listen with an open mind. He became quite unpopular with both Catholics and Protestants, so when he moved to Lyon, he adopted a pseudonym, Michel de Villeneuve.

In 1546, he completed a draft of a treatise he wrote about religion *Christianismi restitution* (On the restitution of Christianity). In it, he opposed baptism of infants as well as the idea of the Trinity. Amazingly, within this 700-page document on religion, Serveto describes pulmonary circulation; this is the first time it was correctly described by a European physician. Serveto wrote that he believed that an understanding of the movement of the blood would lead to a greater understanding of God. He recognized that Galen's system was not correct, because by Serveto's observation the blood seemed to travel to the lungs for its own nourishment, a point that Galen did not realize. Serveto noted that the pulmonary *artery* was very large and that blood moved forcefully from the heart to the lungs, so he considered that more blood than was necessary to nourish the lungs was traveling there and that there must be a reason for this. Serveto developed the theory that the reason for the change in the color of the blood was because aeration took place—that the bright red blood was charged with air before traveling to the left ventricle. Serveto also concluded that the passages between halves of the heart, written about by Galen, did not exist.

To Serveto, the significance of this treatise lay in the religious ideas he expressed. He sent a draft off to John Calvin (1509–64), a French Protestant reformer who was building a powerful following for a new religious system that taught *predestination.* Calvin corresponded with him a few times, kept the manuscript, and then refused further contact. The Protestant reformers saw Serveto with his very Christcentric view of the world as a dangerous radical. When Serveto could not retrieve his manuscript, he rewrote the whole thing, and arranged for the printing of 1,000 copies in 1553. He then turned against Calvin, openly criticizing him.

The concept of religious freedom did not really exist in Serveto's time. Some of Serveto's letters to Calvin were found and turned over to leaders of the Catholic Inquisition, which was dedicated to rooting out any sort of disloyalty to the church. Serveto was imprisoned, but he managed to escape. Four months later, he attended a lecture given by John Calvin in Geneva, and he was recognized, arrested, and sentenced to death for heresy. He was burned at the stake, and most copies of his writings were destroyed as well.

Later, it was discovered that three copies of Serveto's works had survived but had been hidden, and as a result pulmonary circulation continued to be largely misunderstood. It was left to William Harvey to more fully express this theory. (See chapter 4.)

REALDO COLOMBO FURTHER ILLUMINATES THE BLOOD

Vesalius's anatomical studies were later pursued by Realdo Colombo (ca. 1516–59), an Italian *apothecary* who became an anatomist and laid the foundation for William Harvey to eventually explain the flow of blood.

Colombo apprenticed to a well-respected Venetian surgeon for seven years and went on to study surgery and anatomy at the University of Padua. In 1543, Vesalius, a professor at Padua, left to oversee publication of *Fabrica,* and Colombo took over the teaching position he vacated. Colombo eventually moved on to become the first professor of anatomy at the University of Pisa. Later, he moved to the Papal University in Rome where he became surgeon to Pope Julius III.

Colombo was particularly skilled at dissection, and as he worked he began to realize that Vesalius was in error about the passage of blood within the heart. He noted the structure of the vessels, the absence of pores in the septum, and the location of the vessels. He obtained fetuses to dissect and noted that some vessels seemed to circle around the lungs. He outlined the circulation of the venous blood from the right ventricle through the pulmonary artery to the lungs, where it emerges bright red after mixing with "spirit" in the aria, and returning to the left ventricle through the pulmonary vein. He noted that the pulmonary veins had blood, not air (pneuma) as Galen had taught. He also described the general action of the heart, stating that the blood is received into the ventricles during *diastole* (relaxation) and expelled from them during *systole* (contraction). His work on living animals and human cadavers gave him good insight on anatomy, and he wrote well and accurately about the organs within the thoracic cavity, including the *pleura* (*membrane* surrounding the lungs) and the *peritoneum* (membrane surrounding the abdominal organs).

Colombo may have defined pulmonary circulation as early as 1545, but his work *De re anatomica* (On things anatomical) was not published until 1559 when his children made certain that it happened. It was highly critical of Vesalius's work and contained Colombo's theories of the movement of the blood within the body. (He may have read Miguel Serveto, and it is not clear how much of Ibn an-Nafis's theories were known to the Italians.)

Colombo was the first well-known anatomist to write on pulmonary circulation. Even then, his reputation was not strong enough to overcome the power of Galen's writings. It took another 70 years before William Harvey came along and made public headway in this area.

FALLOPPIO AND HIS DISCOVERIES

Gabriele Falloppio (1523–62), often referred to by his Latin name Fallopius, was an Italian anatomist who served as professor of surgery and anatomy at Pisa (1548–51) and Padua (1551–62). While he is associated with the discovery of the *fallopian tubes* (the oviducts that extend from the ovaries to the uterus), his primary focus was on the anatomy of the head. Botany was another of his interests, and he made significant contributions to the medicinal use of plants.

Falloppio was born into a very poor family in Modena, and gaining an education was a struggle. Since clerics had access to education, Falloppio became a member of the religious order at Modena's cathedral in 1542 and as a result was able to study medicine at one of the best schools in Europe. In 1548, he received his

Fallopian Tubes and the Female Reproductive Organs

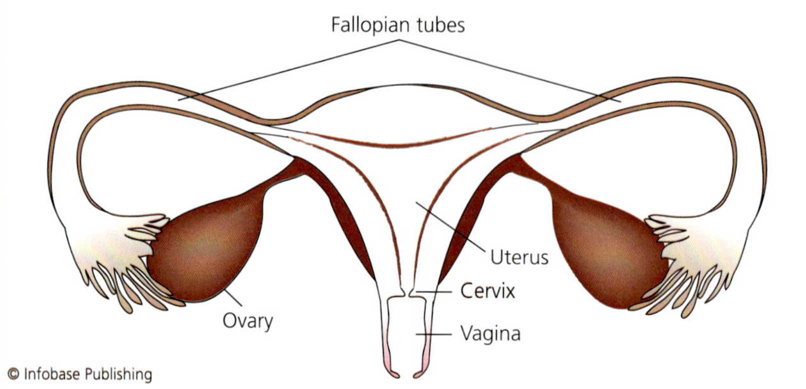

Gabriele Falloppio studied many parts of the human anatomy, but his contributions to the understanding of the female reproductive organs may be the best remembered.

degree from the university in Ferrarra, Italy, and soon became a professor of anatomy. In 1551, he transferred to be professor of anatomy and surgery at the University of Padua. He also supervised the botany department, so his knowledge of medicinal plants grew. He became interested in various therapies and wrote one particular treatise on the benefits of baths and thermal waters. Another treatise focused on the use of *purgatives,* and still another talked about the compositions of various medicines.

Falloppio's primary focus was on the anatomy of the head. He studied the internal structure of the ear, the semicircular canals of the inner ear (responsible for maintaining body equilibrium), describing the tympanum and something about how it worked, and he examined and wrote about the *cochlea* as well as the mastoid cells and the middle ear. He noted the lachrymal passages of the eye and the ethmoid bone and its cells in the nose. His study of the muscles was particularly notable. He was the first person to use an aural speculum for examining the internal parts of the ear.

In addition to the oviducts (now known as the fallopian tubes), he identified other parts of the female reproductive system, including the vagina and clitoris, and noted the existence of the placenta during birth. These anatomical observations were vital to understanding the female reproductive system, but two more centuries passed before scientists began to understand how the eggs traveled from the ovaries to the uterus via the fallopian tubes. He was regarded as an authority on sexuality for his day, and in his writings about syphilis he noted the importance of condoms. (See chapter 6.)

He published only one book during his lifetime, *Observationes anatomicae* (1561), and in it he joined Vesalius in an assault on Galen's theories. Because he was well regarded as a physician and surgeon as well as a scholar, Falloppio's ideas lived on via manuscripts of his lectures, and about a dozen years after his death they were finally published.

BARTOLOMEO EUSTACHIO: FOUNDER OF MODERN ANATOMY

Bartolomeo Eustachio (1520–74) was an Italian anatomist who is now considered one of the founders of modern anatomy. Eustachio's place in history would have been in the same rank as Vesalius if his work had not been misplaced. Only eight of his 47 engraved copper plates of anatomy were located immediately after his death. Had his works been fully published during his lifetime, his discoveries about human anatomy could have helped science in the 1550s instead of 150 years later.

Eustachio was among the students who benefited from the change in church laws (and sentiments) that occurred in 1537 when permission for human dissections in anatomy classes was given. Students from that time forward, including Eustachio, were among the first to have relatively easy access to fresh cadavers for dissections.

Eustachio was born in a small town in eastern Italy. His father was a physician, and Eustachio received a classical education that included the study of Greek, Hebrew, and Arabic. He studied to be a physician at the Archiginnasio della Sapienza in Rome and began practicing medicine around 1540. In 1547, he became the

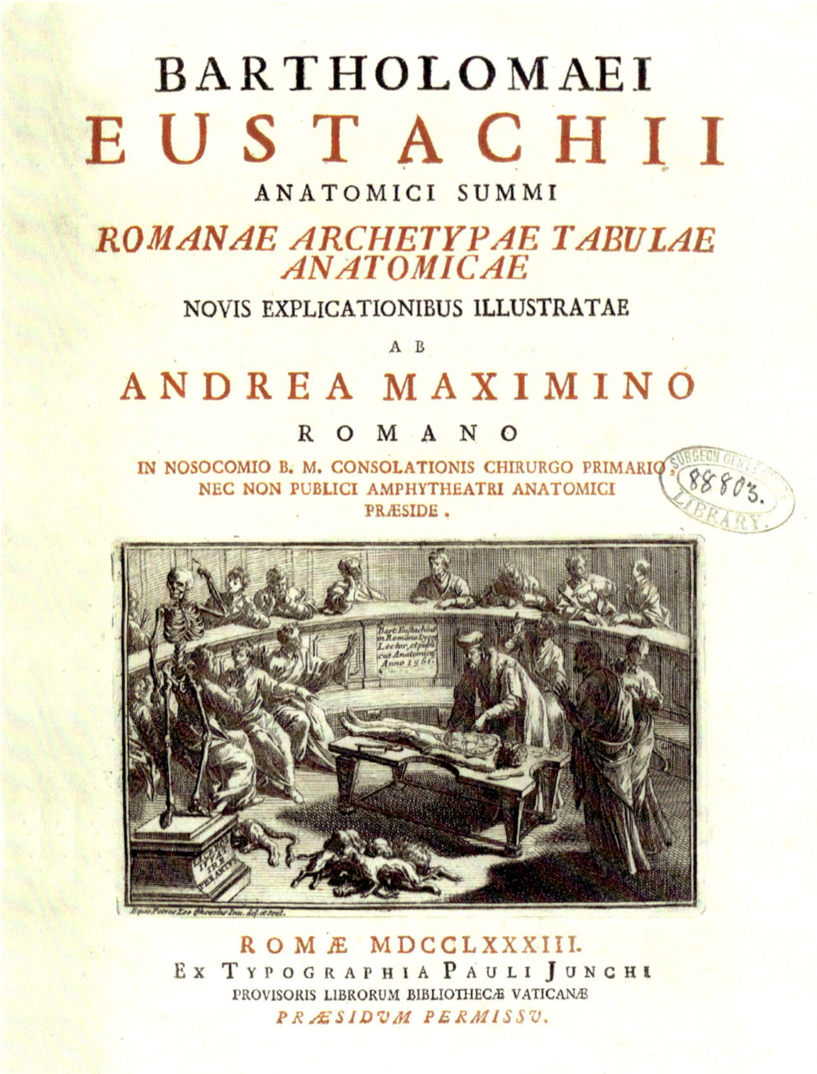
BARTHOLOMAEI
EUSTACHII
ANATOMICI SUMMI
ROMANAE ARCHETYPAE TABULAE ANATOMICAE
NOVIS EXPLICATIONIBUS ILLUSTRATAE
AB
ANDREA MAXIMINO
ROMANO
IN NOSOCOMIO B. M. CONSOLATIONIS CHIRURGO PRIMARIO
NEC NON PUBLICI AMPHYTHEATRI ANATOMICI
PRÆSIDE.

ROMÆ MDCCLXXXIII.
Ex Typographia Pauli Junchi
PROVISORIS LIBRORUM BIBLIOTHECÆ VATICANÆ
PRÆSIDUM PERMISSU.

Note the classroom dissection depicted in the picture. *(National Library of Medicine, History of Medicine)*

physician to Cardinal Giulio della Rovere and also professor of anatomy at the Archiginnasio della Sapienza.

With access to human cadavers, Eustachio began pointing out that previous dissections involving animals bore little relation to human anatomy. Starting in 1552, Eustachio and Pier Matteo Pini, a relative who was an excellent artist, began to work together. They created a series of 47 engraved copper plates based on Eustachio's observations during his dissections. Only eight of these works were published during Eustachio's lifetime, in 1564 in *Opuscula anatomica,* and they provided excellent studies of the kidneys, heart, veins in the arm, the ear (and related elements of hearing), the mouth, and teeth. He wrote an entire book dedicated to the kidney, *De renum structura.*

How the Ear Works

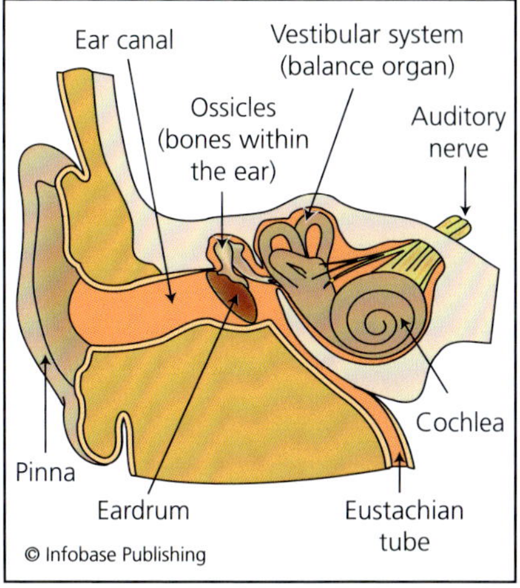

Eustachio was instrumental in beginning to understand the anatomy of the head, including the workings of the ear.

Eustachio also undertook dissecting cadavers of fetuses and newborn babies, and he particularly noted the difference in the mouth and teeth when comparing infants and adults. He wrote about this in *De dentibus* and described the number of teeth in babies and adults and reported on the soft and hard parts of the mouth.

Eustachio eventually retired from teaching because he suffered from bad bouts of gout. He maintained his attendance to Cardinal Rovere and died on his way to check on the cardinal at the cardinal's country home.

Eustachio's contemporaries knew that only part of his work had been published, and they attempted to locate the other 39 plates but were unsuccessful at doing so. Then in the early 1700s, the engravings were discovered by a descendant of Pier Matteo Pini, the artist who had helped him. Eustachio had given him the plates, and 150 years later they were found among the descendant's belongings. Pope Clement XI (1649–1721) purchased them and gave them to his physician who oversaw their publication. The plates provided excellent descriptions of the base of the brain, the sympathetic nervous system (the nerves that control the constriction of blood vessels, among other things), the vascular system, and the structure of the larynx.

SANTORIO AND THE BODY AS MACHINE

The history of the scientific study of metabolism spans several centuries and has moved from examining whole animals in early studies to examining individual metabolic reactions in modern biochemistry. Santorio Santorio (1561–1636) got it started. He was an Italian physician who helped the medical profession into a world of greater precision. A friend of Galileo's, Santorio adapted some of Galileo's inventions for use in medicine; one of the devices was a pulse clock (1602) and another was a thermometer for clinical use (1612). He also invented a device he called a "pulsilogium," which measured the pulse. This was the first machine to do so. A century later, another physician de la Croix used the pulsilogium to test cardiac function. Santorio's prime work and biggest contribution was that he created the first systematic method for studying metabolism. (*Metabolism* comes from the Greek word for "change.")

The concept that the body needed to be continually nourished dates to Islamic physician Ibn an-Nafis who noted that the body was continuously undergoing change as it altered from dissolution to gaining nourishment. The first controlled studies of the metabolic process in humans were undertaken by Santorio. He saw the body as a machine and became interested in studying weight and its relation to food intake.

Santorio created a *steelyard balance* that he could sit in. Over a 30-year period, he studied himself carefully. He described how he weighed himself before and after eating, sleeping, working, sex, fasting, drinking, and excreting. In his book *De statica medicina* (On medical measurement, 1614), he found that the sum total of visible *excreta* was less than the amount of substance he ingested,

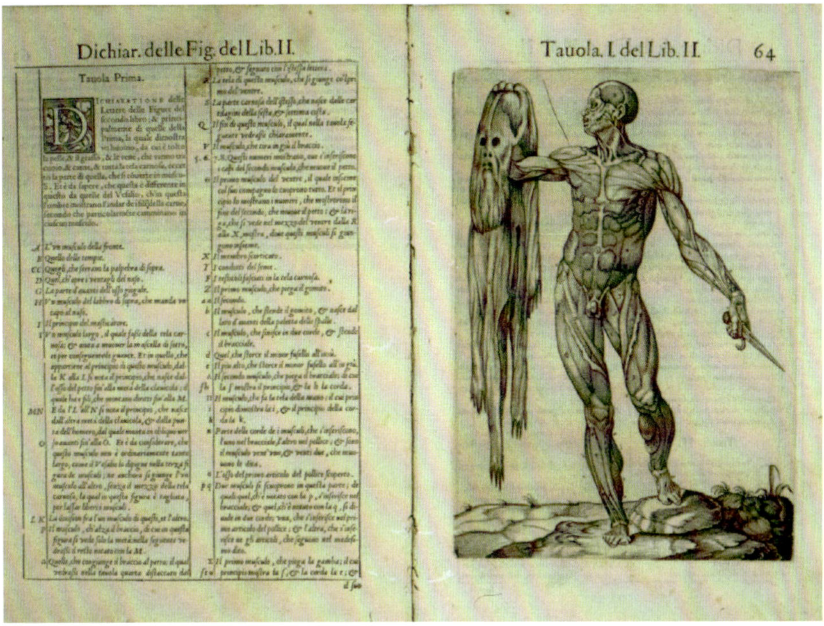

Valverde was one of a group of anatomists who worked in Rome in the middle years of the 16th century. Here, a muscle man holds up his own flayed skin; the accompanying text points out the independence of the illustration from that of the pioneer Andreas Vesalius and discusses Valverde's differences with Vesalius's teaching. *(Vatican Hall, The Library of Congress)*

and this led him to the conclusion that some of what he ate was lost through what he called "insensible perspiration" as a way to account for the difference. *De statica medicina* went through five editions and was published regularly until 1737.

While his findings ultimately did not have scientific value, his achievements were in the empirical methodology he used. He was one of the first to pay such careful attention to gathering and evaluating data. The big change that occurred in the study of metabolism did not occur until the beginning of the 20th century when Eduard Buchner discovered enzymes. At this point, it was possible to separate the study of the chemical reactions of metabolism from the biological study of cells, and this marked the beginning of biochemistry.

CONCLUSION

For the first time, remarkable strides were being made in discovering the human anatomy. Andreas Vesalius made progress by being willing to differ from Galen. Miguel Serveto's new understanding of pulmonary circulation—while not widespread—helped to increase knowledge, which Realdo Colombo was better able to transmit to others. Gabriele Falloppio made notable advances in studying the skull, the ear, and the female reproductive system, and Bartolomeo Eustachio located the eustachian tubes and important ducts and nerves. Metabolism was not well understood at this time, but Santorio Santorio undertook the study of it, and his knowledge laid the groundwork for others to more fully explore how the human body creates and burns energy.

3

Amazing Advances in Surgery

Surgical procedures between the years 1450 and 1700 were often high-risk procedures, and as in the Middle Ages the work was largely left to barber-surgeons. These men trained directly for surgery and had no background in anatomy or medicine, yet their work required incredible skill and a steadfast personality. Professional guilds for various specialties had become important during the Middle Ages, and by 1540 the Guild of Surgeons merged with the Barbers Company to form the Barber-Surgeons Company. This guild established training procedures, controlled membership, and on the whole increased the professionalism of the members.

The type of surgery undertaken was dictated by necessity as well as predicted outcome, with surgeons preferring to take on only those operations that they thought would end favorably. Practitioners were more likely to perform less invasive procedures such as removal of surface *tumors,* coping with broken or dislocated limbs, repair of knife wounds, tooth pulling, draining abscesses, bloodletting, or treating sores (often one of the symptoms of venereal disease). Kidney stones were so painful that on occasion surgeons would attempt surgery to remove them, and trephining was sometimes undertaken.

Two elements—*anesthetics* and *antiseptics*—were still not available to surgeons and their patients. As a result, the pain of surgery continued to be a major issue. Alcohol or an *opium*-based drink were sometimes administered to dull the pain, but barber-surgeons frequently had no time to administer a palliative drink—and sometimes they simply had no access to anything helpful. Generally, a patient was tied down—or held down—by surgical assistants or family members for the duration of the procedure. Patients frequently "roared out," so surgeons had to have courage and the conviction that they were doing the right thing. Time was of the essence, and a "swift hand, sharp knife, and cool nerve" were primary qualifications for anyone performing surgery.

Infection was a very real problem, and unclean hands and dirty surgical tools were not even understood to be an issue. Samuel Pepys (1633–1703), the noted diarist who recorded much about his life for about a 10-year span, underwent surgery for "bladder stone" when he was 25 (1658). The surgeon had rarely performed the surgery before. He made a three-inch incision, removed the stone, and because the surgeon did not know how to close the wound simply told Pepys to stay in bed for a week and let the cut heal "naturally." Pepys survived, and the surgeon successfully performed several more similar operations, but as time went on, his success rate dwindled, probably because no one realized the necessity—nor had the means—to sterilize the surgical tools in between patients.

This chapter discusses those individuals who helped create new and better ways of performing surgery. French surgeon Ambroise Paré was certainly first among those who made a difference in the field, and his contributions spread more widely than they might have because he wrote in French, not in Latin as was the custom for medical texts of the day. Because the books could be reproduced more affordably because of the printing press, Paré gained a strong following of surgeons—including one *midwife*—who made definite improvements in the way surgery was handled.

THE FATHER OF MODERN SURGERY

Barber-surgeon Ambroise Paré (1510–90) was instrumental in changing the practice of surgery. He was well aware that surgery was risky and only resorted to it when he found it absolutely necessary, but his willingness to experiment coupled with sincere compassion for his patients put Paré in the forefront of change during his time.

Ambroise Paré was born in northwestern France in 1510. Sources disagree as to his family background; his father may have been a country artisan or he may have been a barber-surgeon. Paré apprenticed to a local barber (perhaps his own father), and then at 19 he traveled to Paris and became a surgical student at the well-known hospital, the Hôtel-Dieu. In 1536, he attained the rank of barber-surgeon and joined the army as a regimental surgeon. Over the next 30 years, he returned to military service when he was needed, gaining a fine reputation for his considerate and democratic treatment of soldiers of all ranks.

Ambroise Paré is considered one of the fathers of modern surgery. *(Dibner Library of the History of Science and Technology)*

In 1552, Paré entered royal service under Henri II. When the king received a blow to the head in a joust in 1559, both Paré and Vesalius, the well-known anatomist (see chapter 2), were called in to consult on the case. Much to their puzzlement, Vesalius and Paré noted that Henri had received a blow to the right side of the head, yet his left side was paralyzed. Using the heads of four recently decapitated criminals, the two healers attempted to understand the nature of Henri's injuries to see what could be done for him. They were unable to devise an explanation as to why an injury to one side of the head affected the opposite side of the body, and eventually they realized that the injury was fatal. Despite Henri's death, Paré's reputation was so favorable that he managed to stay in the royalty's good graces. He continued in service to kings until the end of his life, serving Henri II, Francis II, Charles IX, and Henri III.

Paré was very religious and felt he did what he could, but a patient's ultimate fate was left to the will of God. His motto, as inscribed above his chair in the Collège de St-Cosme where he eventually taught, read: "Je le pansay et Dieu le guarist." ("I treat him, God healed him.")

A Great and Unexpected Discovery

Warfare has always been terrible, but during Paré's day soldiers who remained healthy were just plain lucky. With no refrigeration or the ability to mass-produce food, armies were ill-prepared to create mobile communities that could feed and house soldiers, and medical care was seriously misguided and usually carried out under less-than-ideal circumstances. Hippocrates had said, "He who wishes to be a surgeon should go to war." The battlefield has always been the ultimate medical school, and it certainly was for Paré, who felt the work he did in helping the soldiers was a divine calling.

Today, battlefield victims are screened so that priority is given to those who are more seriously hurt and most likely to be saved by early treatment. During Paré's time, care was not managed according to who needed it most; priority was given to wounded officers. They also received what was considered a higher level of care. An officer with an injury would be treated with bleeding, cupping, or sweat-inducing drugs, and a foot solder with a similar injury was more likely to be wrapped in a cloth, covered with hay, and buried in manure up to his neck to encourage sweating, according to Lois N. Magner, the historian and author of *A History of Medicine.* Neither cure would have been effective, but greater effort was certainly given to trying to help the officer.

A CHANGE IN WEAPONRY NECESSITATES A CHANGE IN WOUND CARE

The Chinese created gunpowder as early as the ninth century, but Europeans did not begin using gunpowder until weapons were created that permitted the powder to be fired directionally. Cannons were used by the 14th century and eventually hand-carried weapons came into increasing use. The first such firearm was the *harquebus,* which came into use between the 15th and 16th centuries. These guns had long barrels with a flared end to make them easier to load with the gunpowder, and they were fired by using a matchlock, which had to be lit by a slow-burning match. It could be carried by one soldier, but it had to be braced on a pole with a forked end before firing. After the harquebus, the *musket* was invented, and it offered the decided advantage of being lighter, more accurate, and could be handled by one person. By the 16th and 17th centuries, firearms were being used more frequently throughout Europe, causing a major change in warfare and the medicine needed to treat injuries.

Wounds caused by gunpowder were far more damaging than those inflicted by arrows and swords. When the gunpowder hit the body, it tore into and destroyed a wider area of flesh and human tissue, opening larger wounds that led to deeper and more widespread infection. As physicians and barber-surgeons began to see the new types of injuries, it necessitated changes in the way that the wounds were handled. From the time of Giovanni Vigo (ca. 1460–1520), a surgeon-in-ordinary to Pope Julius II, gunshot wounds began to be classified. Three general categories included contused, burned, and poisoned. Vigo was one of the first surgeons to write about how to handle gunpowder wounds, and he noted the use of boiling oil to neutralize the poison of the gunpowder.

Paré was a surgeon in the army of Francis I from 1536–38 in Turin (now Torino in the Piedmont section of Italy), where the French army was fighting to take over the area. The surgeons were seeing more and more damage from gunpowder. (See previous sidebar.) To stop the bleeding and rid the wound of the gunpowder "poison," surgeons were following the traditional practice of pouring boiling oil on the wound. The pain for the patient was excruciating, frequently sending the soldier into shock. During one battle, Paré treated so many wounds that he ran out of oil and had to improvise. Creating a mixture out of egg yolk, rose oil, and turpentine, he used this "plaster" on those soldiers who still required treatment. That night he was so worried about the fate of his patients that he did not sleep well and arose early to check on how they were doing. He later wrote: "To my surprise I found those to whom I gave my ointment feeling little pain, and their wounds without inflammation or swelling, having rested reasonably well during the night. The others, on whom I used the boiling oil, were feverish with great pain and swelling around the edges of their wounds." As a result of this discovery, his reputation grew, and he was held in

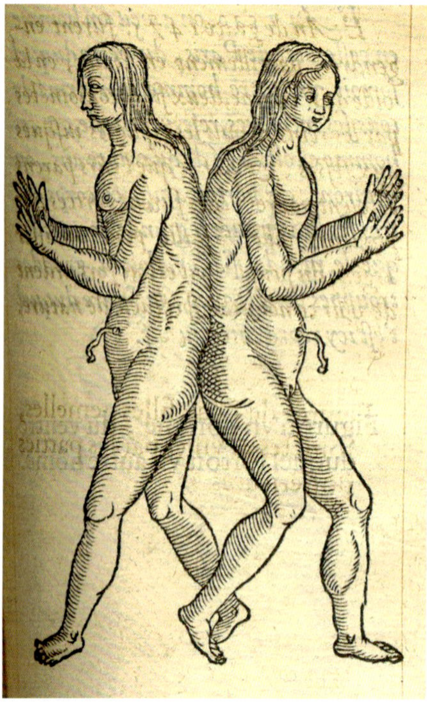

Deux livres de chirurgie. Paris: André Wechel, 1573. Although there were many 16th-century treatises on monsters, surgeon Ambroise Paré brought a more scientific approach to the genre in his treatise *Des monstres et prodiges.* This woodcut illustration depicts two unnamed female conjoined twins who lived in Verona, Italy, circa 1475. Paré informs the reader that they were joined at the posterior from the shoulders to the buttocks and shared their kidneys.

special regard by the soldiers. Not only had he begun to learn that tried and true was not always best, but he had inadvertently also created a clinical trial of sorts where he could perform a side-by-side comparison of two different treatments.

He wrote about his experience in *La méthode de traicter les playes faictes par hacquebutes et aultres bastons à feu . . .* (Method of treating wounds made by harquebuses and other guns . . .) in 1545. While Paré's findings bore a lot of validity, the book drew negative attention because Paré had written it in French rather than Latin as was traditional for the writing of medical books. Though Paré's pride was wounded by the criticism, the fact that he wrote in the vernacular made the book much more accessible to barber-surgeons, few of whom would have been able to read a text in Latin.

Over time, Paré contributed greatly to surgery, primarily because he could convey information to a wider audience. He also realized that a better understanding of anatomy was vital to anyone attempting to do surgery. In 1561, he created another work written in French in which he summarized large sections of Vesalius's material on anatomy. By including Vesalius's information, Paré made it possible for barber-surgeons to begin to understand the parts of the body.

What Paré Learned about Amputations

Extensive experience on the battlefield caused Paré to carefully examine the process of limb amputation, and he applied his knowledge and compassion in order to bring about change in the surgical process, the need for substitute limbs, and the patient experience of phantom pain.

Surgically, the amputation process was brutal. There was generally no opportunity for any pain-numbing drink, and the soldier's limb was generally cut off with some form of surgical hacksaw. Next, most surgeons applied a hot iron to the wound to stop the bleeding. This process burned off any skin that might have been used to cover the wound, so the stub of the limb was left open and unprotected.

Paré realized that having a limb cut off was bad enough for a patient, and there needed to be another way to manage an amputation. He rejected the use of *cautery,* which burned off the skin while stopping the bleeding. He realized that if he could save a flap of the skin it could be used to cover the wound. Reverting to a method used by the Greeks, Paré began tying off blood vessels with silk thread to stop the bleeding. This process generally required at least one assistant to help with the tying off as it had to be done quickly before the patient lost too much blood. Though they had no knowledge of the necessity of hand-washing to avoid spreading infection as they tied off the blood vessels, in general, it was an improvement over the cauterization process. Paré's *ligature* technique was described in his book, *Treatise on Surgery* (1564).

Working with so many soldiers who were wounded in battle gave Ambroise Paré a heightened understanding of the issues faced by amputees. Paré listened to amputees talk and noted the phantom pain (sensation in the amputated limb) that amputees experience. He came to believe that phantom pain arose from the brain, and the medical community today agrees with this. Paré also realized the necessity of substitute limbs, and he created artificial legs that could be used by the poor as well as the rich. He also introduced the implantation of teeth and in an acknowledgment that sometimes what is important is looking normal created artificial eyes.

PARÉ IMPLEMENTS MANY ADVANCES

Because Paré experimented and shared his knowledge, he raised the prestige and the level of professionalism of his trade. He was also an innovator in many areas:

- He invented an early *hemostat clamp* (he called it the "crow's beak"); it was a scissorlike tool that could be inserted to apply pressure to stop bleeding. A version of his invention is still used today both in surgery and in emergency medicine.
- Paré continued to use cautery for some circumstances, but he rejected the use of acid treatments to burn the

wound and stop the bleeding. He preferred to use hot irons that he created himself.

- He advanced *obstetrics* by reintroducing a method used to turn a fetus in utero to ease delivery. Successfully turning the baby for a head-first delivery was preferable, but sometimes the best that could be done was getting the feet down for a breech delivery. He also learned to induce labor when necessary.
- Paré was very sensitive to the experiences of his patients. If there was time or opportunity, Paré's patients were given wine, an opium mixture, or henbane to deaden their pain.
- He learned about using chopped raw onions to heal wounds from a female healer, and in the 1950s scientists analyzed this and discovered that onions actually contain an antimicrobial agent.
- He also learned to do a proper herniotomy. In that day, hernia patients' tissues sometimes were strangled by the hernia and patients frequently died. Paré's surgical method saved many of them.

Among his missteps was advocacy of puppy oil, which he claimed was soothing. This required boiling newborn puppies in oil of lilies, mixing the mixture with turpentine and a pound of earthworms. Ultimately, he was able to benefit from his own medical intuition. Late in life, Paré was kicked by a horse, which resulted in two broken bones in his leg. The common treatment at the time for such a debilitating injury would have been amputation, but Paré knew this would put an end to his career. He asked those who were around him to bring him what was available in the village, and they came back with egg whites, wheat flour, oven soot, and melted butter. He splinted his leg and then used the ingredients to create a cast for himself. He also kept adding rose oil to the abscesses and letting them drain. Eventually the leg was well enough for him to get around, and the cast held the bones in place until it more fully healed. He regained use of his leg.

DEBUNKING POPULAR MEDICINES OF THE DAY

Bezoar stones were calcified "stones" from the intestines of animals, and they were considered an *antidote* to poisons as well as the cure for a variety of chronic and painful diseases. Sometimes the stone was pulverized and added to a drink or steeped in boiling water for a time, giving the patient the resulting brew. Other times it was sucked on. Some healers taught that the Bezoar stone should be set in a ring (from which setting it could be removed). This made it convenient to carry along to suck on several times a day. This was expected to cause profuse sweating that resulted in an antidote to poison. Sometimes patients used an emetic or other type of medicine to clean out the body first and then for several successive mornings drank a preparation made from the stone.

Paré did not believe that the Bezoar stone had these curative properties, and when an opportunity to test his theory presented itself he seized it. A cook who worked in the kitchen of Charles IX, one of the kings with whom Paré was associated, had been caught stealing two fine silver plates and was sentenced to hang for his misdeeds. Paré offered him a deal:

OTHER NOTABLES IN THE FIELD OF SURGERY

While Paré was a leader in his field, the great need for surgical expertise meant that others began exploring the field. The following are among those who made progress.

Thomas Gale Crusades against Charlatans

Thomas Gale (1507–87) was a British surgeon who put in long service on battlefields, and his experiences led him to actively crusade against charlatans. He was particularly struck by the battle

Agree to take a poison, and Paré would make certain that he also received a drink brewed from the Bezoar stone. If the Bezoar stone worked and the fellow survived, he would be given his freedom. If not, he would meet with the same fate he was going to meet with anyway. Unfortunately for the cook, Paré was correct. The stone was not curative, and the cook died an agonizing death seven hours after taking the poison.

Paré also debunked other medicines that were popular in his day. Physicians thought that powder from unicorn horns and from ancient mummies were medicinal. Because unicorns were so rare (actually mythological), people substituted narwhal and rhino horns for the unicorn horn. Paré ran several tests of the effect of the powder on spiders, toads, and pigeons. He found no helpful cure from it. Apothecaries and physicians were annoyed by Paré and retorted that he was simply using "cheap" substances . . . that with the right ingredients, more expensive ones, a cure could be realized. Paré had the perfect response. He said he would rather be right and stand alone than stand in a group and be wrong.

aftermath he noted at Montreuil in 1544. He observed that soldiers were being treated by tinkers and cobblers who claimed surgical knowledge but were actually making things much worse for the soldiers. Gale wrote that these charlatans were dressing the wounds with mixtures including ingredients such as the grease of shoemakers' wax and rust from old tea kettles. Even those with minor wounds ended up dying.

Later on, Gale was helping out at St Thomas' and St Bartholomew's Hospitals in London, two hospitals for the poor.

Surgical Tools of the Day

Early surgeons often devised their own tools, and this style of tool was probably used to separate the tissue for further surgical exploration.

According to William John Bishop in *The Early History of Surgery,* Gale noted that most of the patients were extremely ill, having come to the hospital only as a last resort. As he went from patient to patient, he asked to whom they had turned for treatment: "All were brought to this mischief by witches, by women, by counterfeit javills [rascals] that take upon them to use the art, not only by robbing them of their money but of their limbs and perpetual health."

When Gale returned from time on the battlefield, he was promoted to serjeant-surgeon to Elizabeth I. Among his writings were *An Excellent Treatise on Wounds Made with Gunneshot* (1563) and *Certain Works of Chirurgie* (1586).

William Clowes: Master of Wound Treatment

William Clowes (1544–1603) was born into a well-off British family. Clowes is thought to have come to London in 1556, at the age of 12, to begin what was usually a seven-year apprenticeship in becoming a surgeon. He was apprenticed to George Keble, who was well thought of and also practiced physick to help make people better. Clowes adopted many of the ointments, plasters, and prescriptions that he learned from Keble.

When his apprenticeship ended, he served the Earl of Warwick's army that was fighting in Normandy in 1563. While serving the soldiers in the military campaign, Clowes developed a lifelong friendship with John Banester, a fellow surgeon. Warwick's army returned to England a year later, but Clowes remained at Portsmouth to serve the sailors who needed help.

Over time, he became one of the most experienced surgeons in treating men in active service. He was eventually to write about treatment of these wounds in *A Prooved Practice for all young Chirugeons, concerning burning with gunpowder, and woundes made with Gunshot, Sword, Halbard, Pike, Launce or such other* (1588). He also wrote a short book on the treatment of syphilis, another topic on which he had gathered knowledge.

In 1576, he became a surgeon at Christ's Hospital and eventually in 1581 went on to be a full surgeon at St Bartholomew's Hospital, but interrupted his tenure there to go to the Low Countries to attend to her majesty's forces on the battlefield. In 1588, he left battlefield service to be "one of the Queen's Chirurgeons," where he was able to lecture and teach as well as serving the queen. He ended his career in private practice from his country home in Essex.

Tagliacozzi: Revived Indian Methods of Plastic Surgery

Gaspare Tagliacozzi (1546–99) was an Italian surgeon who studied at the University of Bologna, earned degrees in both philosophy and medicine, and began teaching there, first as a professor of surgery and later appointed to serve as professor of anatomy at Bologna. In his surgical work he was known as an innovator, and he revived the art of rhinoplasty that was first known to be used by Indian healers. This form of plastic surgery for the nose (described in *Early Civilizations,* the first volume in the History of Medicine series) involved twisting a flap of skin from the forehead to come down over the nose in order to repair or change the nose. As late as 1550, there were descriptions of a similar method that involved removing a flap of skin from the arm to cover the nose. This was a less desirable method as the arm where the skin was removed then had to be immobilized in order to allow healing. Tagliacozzi likely learned the technique from eastern healers who had followed the trade routes from India to Italy.

His principal writings were completed in 1597 in the work entitled *De curtorum chirurgia per insitionem libri duo.*

Richard Wiseman: The Importance of Adaptation

Richard Wiseman (ca. 1623–86) was a contemporary of the respected Thomas Sydenham, often known as the English Hippocrates (see chapter 7). Wiseman was considered one of the greatest surgeons of the 17th century, and, while he had great respect for advances made by Ambroise Paré, he primarily worked on battlefields and had to adapt Paré's methods and create shortcuts in order to save lives. He pointed out that Paré's recommendation to sew a wound closed was admirable, but he noted that when on shipboard with a rolling sea beneath, a cautery was far more efficient than working with a needle and thread.

Fixing a Dislocated Shoulder

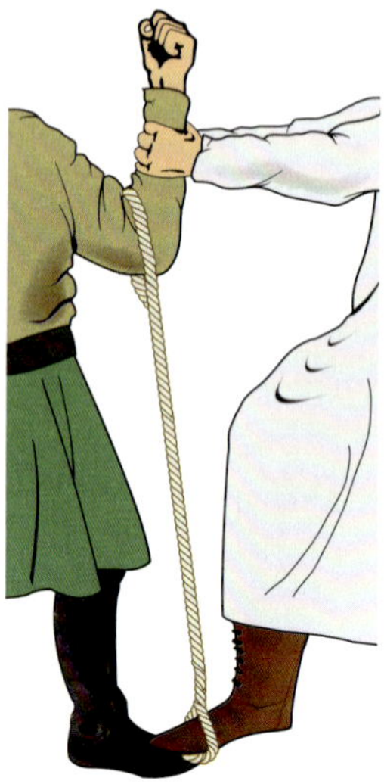

Health issues—such as a dislocated shoulder—required solutions, and these often involved brute force. The fellow on the right provides opposing force by yanking the shoulder while using the rope to stabilize the arm.

Not a lot is known about his early years, but Wiseman became a surgical apprentice at the age of 15 and soon a naval surgeon in the Dutch navy (the Netherlands were favored allies of England). In 1645, he returned to help on the battlefields during the English Civil Wars. After the battle of Worcester, Wiseman was captured and imprisoned, but during that time he was permitted to practice surgery and later on given enough liberty to see patients outside the prison. When he was finally freed, he left England and joined the Spanish navy. When Charles II returned to London in 1660, he asked Wiseman to return. Wiseman was appointed "Surgeon in

Ordinary to the Person," and eventually "sergeant-surgeon and principal surgeon to the King" (1672).

His first book was written in 1672 and was intended primarily for naval surgeons. In it he addressed methods of treating gunshot wounds, fractures, and venereal diseases. With open wounds, Wiseman stressed the importance of removing all foreign bodies before the first dressing, noting that even if the wound needed to be enlarged to accomplish that, it was important to do. (His treatment method for gunshot wounds was actually very similar to the method used 200 years later during the American Civil War; there were few advances in wound care during this time.)

This book on surgery was so well received that in 1676 he brought out an expanded version, *Several Chirurgical Treatises.* It became the authoritative book on surgery and was republished regularly over the next 60 years. The format of the book consisted of a treatise on a particular condition followed by its definition, cause, signs, prognostics, and cure. Wiseman wrote clearly and must have kept copious notes on each patient as his case histories are extraordinarily detailed. He documented both successes and failures because he noted that others should learn from what he did wrong. By incorporating Paré's ideas into his work and his writings, Wiseman was helpful in spreading the French surgeon's ideas to a new audience in England.

Wiseman Notes One "Cure" Left to the King

During this period a form of tuberculosis that involved the swelling of the lymph glands, particularly those in the neck, was known as the king's evil (also known as scrofula). It was believed that the best cure for this illness was the touch of the king. Charles II (1630–85) frequently held public audiences where those who suffered from the disease could see him. These events were actually quite costly for the king, as it was believed that in addition to his touch, he needed to bestow a gold piece to help bring about a cure for each person who suffered.

Wiseman wrote of the king's evil in his book, and he noted the power of the king to heal, describing it as a "miraculous Cure." Of course, to write otherwise would have been considered disloyal.

His book contains other cures for the king's evil, noting that many people do not have this opportunity to avail themselves of the "easy and short remedy." He notes that the tumors seem to arise from a peculiar acidity of the blood serum, and he recommends diet and air as part of the cure, as well as plucking out the lesion when possible.

MIDWIFERY IS IMPROVED

Louyse Bourgeois (1563–1636) was an influential midwife who increased the level of professionalism among those who oversaw the birthing process. Bourgeois likened midwifery to being a ship's pilot—to work with natural forces rather than becoming ensnared in a futile quest to overpower them. Her ethical precepts are still viewed as dominant today. Her story is an interesting one because it highlights attitudes of the era.

Not a great deal about her early years is known, but it is felt that she was born into the middle class because she was taught to read and write in French, not Latin, which would have been taught to daughters of noblemen. Bourgeois married Martin Boursier, an army surgeon and barber, who had studied medicine under Paré. Based on Bourgeois's level of knowledge about medicine, it is speculated that Boursier shared a great deal of what he learned from Paré with her.

Together the couple had three children. In the late 16th century, religious wars were ongoing in France, and Boursier was often off treating soldiers. When the fighting came too near their home in 1589, Bourgeois and her children fled and resettled, with Bourgeois taking in needlework to support the family. A midwife who had attended Bourgeois during the birth of one of her children told her that if one had the ability to read and write and were to learn midwifery great progress could be made in helping women. Though most women learned about childbirth simply by passing information on orally, Bourgeois began to seek out what had been written about childbirth and asking questions of her husband. By 1593 or 1594, she was attending the births of the working-class women in her neighborhood, and both her knowledge and her reputation grew.

The only midwives permitted to help with noblewomen or royalty were an elite group of midwives who were certified by the city (there were just 60 of them listed in Paris in 1601). Bourgeois studied and applied for certification, which involved submitting references and being examined by a panel that included a physician, two surgeons, and two certified midwives. She achieved her certification on November 12, 1598, and almost immediately her services were in demand.

Forceps

© Infobase Publishing

This type of tool was not yet used for childbirth, but it was helpful in grasping anything that was a little out of reach.

In 1601, Henry IV had married Marie, the daughter of a scion of the wealthy de Médicis family, and because royal offspring were highly valued royal pregnancies were very important. At that time, Henry had not yet produced a male heir. Henry recommended that Marie use one midwife, but because that midwife had delivered several of his illegitimate children, Marie wanted someone different, and Louyse Bourgeois was highly recommended. The first child Bourgeois delivered for Marie and Henry was the badly wanted male heir, and she went on to deliver five more children for Marie. She was well paid, earning about 900 livres for each delivery (as opposed to 50, the normal midwife payment), plus a bonus of 6,000 livres in 1608. In 1606, she was given the official title of midwife to the queen, which greatly increased her desirability.

In 1609, she began to publish her knowledge of childbirth, and her book was one of the first treatises on midwifery ever written.

It was filled with more practical information than other books, and it was soon translated into Latin, German, Dutch, and English and relied upon for at least 100 years, until the early 1700s. She eventually produced two more books as well, making her work a three-volume manual on midwifery. She wrote on obstetrics and addressed podalic version (turning a baby), and, as a result, this procedure became widely known within the profession. A review of her writings has led to the conclusion that she was the first to administer small doses of iron to treat *anemia.*

Bourgeois also fought to provide more training for midwives, and in 1635 she sought permission to teach a course on midwifery at the Hôtel-Dieu. When she was rejected, she set up to teach privately, and one of her students eventually became head midwife at the Hôtel-Dieu, where she implemented a method of training. As a result of her influence, midwives in Paris took training and began to follow her example of improving training and attending lectures.

Bourgeois's career seemed to have come to a halt in 1627 when Marie de Bourbon, wife of the brother of King Louis XIII, died in childbirth. The autopsy attributed the death to an infection from a bit of the placenta that had remained inside the uterus. The report did not blame Bourgeois, but she launched a written attack critical of the autopsy panel, and the panel as a group retaliated with a written response of their own. Though she did not die until 1636, it seemed that her career as a midwife ended with this unpleasantness. Several of her children continued in medicine and midwifery.

SURGERY ACHIEVES GREATER RESPECT

Charles-Françoise Félix (1635?–1703) brought the profession of surgery to a new level by healing Louis XIV (1638–1715) of France. Anything suffered by royalty was deemed far more serious than the ills of the common man, and therefore the cures also became that much more important.

Painful cysts sometimes develop in the lower back area, brought about by irritation from activities such as riding horseback (common to medieval knights as discussed in *The Middle Ages,* a previous volume in the series History of Medicine) or riding in carriages on bumpy roads. When Louis XIV developed a cyst, physicians pressed ahead with state of the art treatments of the time. They performed bloodletting with leeches and gave medicines to purge the king's body, but nothing was helping. Though surgery was always a last resort, they finally called in Charles-Françoise Félix, the best surgeon of the day. In *The Illustrated History of Surgery,* the author Knut Haeger notes that Félix realized he did not know enough about this particular condition to treat royalty. Félix contacted hospitals for the poor and located several people with similar issues in order to experiment on them before treating the king. First, he explored nonsurgical methods, since even for surgeons, surgery was a last resort. He suggested different cures ranging from drinking sulphur waters to applying special salves, and eventually he created his own narrow-bladed knife for the surgery. He practiced the operation on several patients; some died from the process. Since nothing else helped, Félix became convinced that surgery offered the best hope for a cure.

As was common with procedures concerning royalty, everything was recorded, and according to *The Illustrated History of Surgery,* the operation took place at 7 A.M. on November 18, 1686, in the king's bedchamber at Versailles. In addition to Félix, three other doctors and four apothecaries were in attendance. Additional people in the room included Louis's new wife, a war minister, a priest, and a secretary to record all that transpired. The records show that Félix cut twice with the knife and used scissors in the wound area eight times. It was noted that the king "never flinched or utter a sound of pain. . . ." However, Haeger notes that the secretary may simply have written this, bearing in mind that the king might read it later and would want to be portrayed as heroic.

An hour later, Louis submitted to a bloodletting. It was noted that the king generally refused this procedure, so speculation was that he was worried or not feeling well. But that evening he held council and the next day he received ambassadors, though he was reportedly in pain. Félix believed that the best way to continue the cure was to keep the wounds from healing too quickly, so he reopened them on December 6, 8, and 10. On January 11, the king finally was well enough for a promenade in the Orangerie at Versailles. The people sent good wishes, and there was a public feast with 236 courses.

As a result of Louis XIV's successful cure, the practice of surgery was elevated.

CONCLUSION

While most practitioners before this time had been reluctant to do much surgery because of high fatality rates, the urgency and necessity of dealing with an expanded number of wounded soldiers injured with gunpowder on the battlefields created opportunities for advances in the field. Ambroise Paré's work was exemplary, but he was soon followed by Thomas Gale, William Clowes, and Richard Wiseman, all of whom contributed knowledge to what was a very young field. Childbirth also provided a regular opportunity for education, and while friends and servants tended to most women giving birth, Louyse Bourgeois was important for illustrating that proper training could make a difference.

4

William Harvey Transforms Understanding of the Circulatory System

Other scientists before him had begun to explore the possibility that the blood circulated, but it was William Harvey's persistent and careful methods that brought about new proof and a better understanding of the workings of the heart and the circulation of the blood. Harvey's discoveries were among the most significant in medicine.

Physician William Harvey (1578–1657) lived at a time when the study of anatomy was beginning to dominate all of medicine. Harvey saw that studying anatomy and understanding the placement of various organs and bones were important but only part of the picture. In order to fully understand the human body, he realized the value of studying physiology, how the body works. In undertaking his studies, he came to see that the blood actually circulates throughout the body; it doesn't get "consumed" by the tissues and organs as was the common belief of the day.

Harvey was not alone in making excellent progress in the study of physiology. As Galileo pushed the use of optical lenses to study vistas beyond Earth, Marcello Malpighi did so with the use of an

William Harvey, first to describe the circulation of the blood *(The Yorck Project)*

early microscope. He was key to affirming what Harvey believed, because it was through Malpighi's studies that the *capillaries* were discovered.

This chapter highlights the circulation of the blood as laid out by William Harvey and the discovery of the capillaries through the application of the microscope by Marcello Malpighi. Several other scientists made major contributions to the understanding of the body's physiology, and their work will also be explained.

EARLIER THEORIES OF THE BLOOD (PRE-HARVEY)

Long before the 16th century, various cultures had been curious about the purpose of the blood, how it was made, and where it might possibly go within the body. (The assumption was that it was continually being made and consumed, and that was one of the reasons why bloodletting seemed like a sensible option. If blood was made continuously, there was no reason to worry about removing some of it.)

The early Greeks explored many elements in efforts to better understand how the body worked. As early as the fourth century B.C.E., Aristotle studied human anatomy and located the blood vessels. Then Praxagoras of Cos (fourth century B.C.E.) noted that arteries were different from veins and put forth the view that air (pneuma) circulated through the arteries, while blood circulated through the veins. Two more scientists came along and further completed this picture. Praxagoras's student, Herophilus of Chalcedon (335–280 B.C.E.) reached a different conclusion, believing arteries carried blood not air. He also studied the body's pulse rate and using a "water clock" developed ways to document pulse strength and rhythm. Thirty years later, another gifted medical practitioner, Erasistratus of Ceos (304–250 B.C.E.), decided that blood in the body must move similarly to the way sap moves in trees. He mapped the veins and arteries and concluded that the heart functioned like a pump to move the blood around.

The theories developed by Herophilus and Erasistratus were very advanced for their time, and had other physicians and scientists used the theories introduced by these Greek physicians as stepping-stones toward better understanding of blood circulation, they would have arrived at a more accurate understanding much sooner. As it happened, Herophilus and Erasistratus's enlightened realizations were not taken seriously. Only fragments of their writing survived to be passed on to other scientists, and their ideas were also trounced by Galen, who dominated medicine from the second century onward. Galen was highly critical of Herophilus and Erasistratus for not adhering to Hippocrates' teachings, and Galen himself was developing his own theories about the blood that he wanted others to believe.

Galen worked primarily on animals (though his written materials never specified this), and he determined that there were two types of blood in humans: the fresh and well-nourished blood (dark red) that traveled via the veins to the right auricle (upper chamber) of the heart and then to the right ventricle (lower chamber) where it passed through the septum of the heart to the left side, mixing with arterial blood that had picked up air from the lungs, making it brighter and thinner. He believed that the blood in the veins was created in the liver from "nutritious substances" (food), and its purpose was to nourish the organs and tissues where it was eventually "consumed" by the organs and tissues. The job of the arteries was to take blood from the heart to the brain where impurities were filtered out and discharged. Galen also taught that the blood moved because the arterial system could contract, causing the blood to ebb and flow like the sea. Galen also identified the vascular network rete mirabile that he said was in the neck of all living things. (This network does not exist in humans, so this was one of the prime ways that scientists came to understand that Galen had not studied human bodies.)

AN ISLAMIC PHYSICIAN PROVIDES OTHER ANSWERS

As early as the 13th century, Islamic physicians had developed a better understanding of how the heart and circulatory system worked. The Western world, however, was unaware of these gains until a 20th-century discovery. In 1924, an Egyptian physician Dr. Muhyi ad-Din at-Tatawi wrote his thesis on some little-known writings of physician Ibn an-Nafis (1210–80). About 40 years later, this doctoral thesis came to the attention of the historian Max Meyerhof, who read the thesis to learn what Ibn an-Nafis believed. Though no contemporary writers of his time seemed to have picked up on Ibn an-Nafis's findings (though some may who have not yet been translated), Ibn an-Nafis had come to understand how the blood travels through the body. Acknowledging Galen's theory, Ibn an-Nafis agreed with Galen that the left ventricle contained vital spirit while the right ventricle contained blood, but he disagreed with Galen's theory about the pores within the septum permitting blood and spirit to pass between the left and right sides of the heart. Ibn an-Nafis theorized that the blood needed to go from the right ventricle to the lungs to acquire air, and only then would it enter the left ventricle. This theory was correct, and it preceded what was learned later by those in the 16th century who were studying anatomy. Andreas Vesalius (see chapter 2) was the first to raise concern that the septum was too dense to permit blood to pass through, and one of his assistants Realdo Colombo of Cremona developed a theory concerning a pulmonary circulation system, which was further developed by his pupil Andreas Casalpinus. Even with these breakthroughs, however, these men still felt that the veins were the key to distributing blood through the body.

As it happened, Galen's theories lived on for the next 1,500 years, because no one else had any ideas that were better (and because Ibn an-Nafis's work was not translated until much later).

Throughout the Middle Ages, human dissections were still frowned upon, and if physicians did not accept Galen's ideas, then they were left with other questions: If the liver did not create blood, what did the liver do? If food was not converted into blood in the liver, then where did the food go and what purpose did it serve? How were the tissues nourished if the tissues did not "consume" the blood?

HARVEY BREAKS NEW GROUND

William Harvey (1578–1657) was the eldest of nine children born into a family in Folkestone, England (in the southeastern part of the country). His father was a successful businessman, and his brothers followed their father into the world of business. Harvey went into medicine. He studied at The King's School in Canterbury and at Gonville and Caius College in Cambridge, from which he received a B.A. in 1597. He then continued to the University of Padua where he studied under the well-respected anatomist Hieronymus Fabricius, graduating in 1602. His marriage to the daughter of a prominent London physician helped him with connections in the medical world, and he soon was given a position at St Bartholomew's Hospital and also served as a fellow of the Royal College of Physicians.

During the course of his career, he was physician to James I (James VI of Scotland, 1566–1625), who succeeded Elizabeth on the throne of England in 1603, and he was physician to Charles I when he took over in 1625. Harvey was promoted to be "physician in ordinary" (the title of the highest-ranking physician in royal service) shortly after.

New Discoveries

Harvey's writings show a man who admired Aristotle and valued the views of Galen. But he also relied on his own observations and reasoning to develop his conclusions. While studying under Fabricius at the University of Padua, Harvey benefited from Fabricius's discovery of "valves" within the veins. (Fabricius wrote *On the*

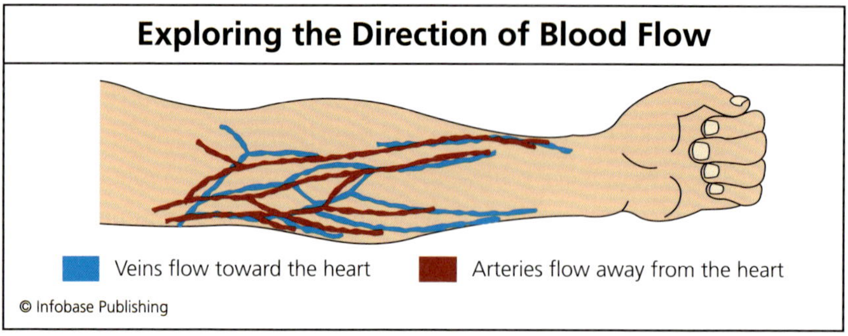

William Harvey began to understand that veins and arteries served different purposes.

Valves of the Veins, published in 1603.) Harvey was fascinated by this theory, but was still puzzled by the purpose of the veins and did not feel that Fabricius had properly explained the purpose of the veins or how they worked.

From Fabricius, Harvey had learned of the value of comparative anatomy, so he began to dissect all types of things—from insects and earthworms to reptiles, birds, and mammals, as well as human cadavers when he had access to them. He particularly wanted to examine the heart and the movement of the blood, but he found that in warm-blooded animals the systole (contraction) and diastole (expansion) happened so rapidly that it was hard to observe what was happening. He soon realized that by conducting *vivisections* on cold-blooded animals, such as snakes and frogs and fish, he could observe the heart better because hearts move more slowly in cold-blooded animals, making them easier to study.

As he experimented, he came to realize that the veins seemed to carry blood in one direction only and that was toward the heart. To prove this, Harvey devised a method to test what he believed. He placed a ligature on a person's upper arm to cut off blood flow both from the arteries and veins; he then noted that the arm below the ligature was cool and pale, while above the ligature it was warm and swollen. By loosening the ligature, he witnessed the change in blood flow. He also saw that he could push blood in the vein up toward the heart but there was no way to push it downward—the veins only moved blood in one direction. Harvey came to understand that the bumps in the veins were the valves discovered by his teacher, Fabricius, and they were the devices that maintained the one-way flow.

Next Harvey concluded that the blood moved because the heart was muscular. He showed that blood was expelled from the ventricles during contraction or systole (and was sent out through the body) and flowed into them from the auricles during expansion or diastole. He proved that the arterial pulse was due to passive filling of the arteries to the systole of the heart and not by active contraction of the walls. As early as 1603, Harvey wrote "the movement of the blood occurs constantly in a circular manner and is a result of the beating of the heart." He also noted that the blood seemed to flow in two closed loops. One (the pulmonary system) took the blood into the right side of the heart from which it passed into the lungs before going into the left side of the heart; while the other sent the blood out to the rest of the body.

Harvey also undertook the first "quantitative" studies of the blood, measuring how much blood passed through the heart each day. He worked with estimates of the capacity of the heart, tried to measure how much blood was expelled with each beat, and counted the number of times the heart beat in half an hour. Based on the information gathered through these studies, Harvey realized that the liver could not be producing the blood with the body consuming all

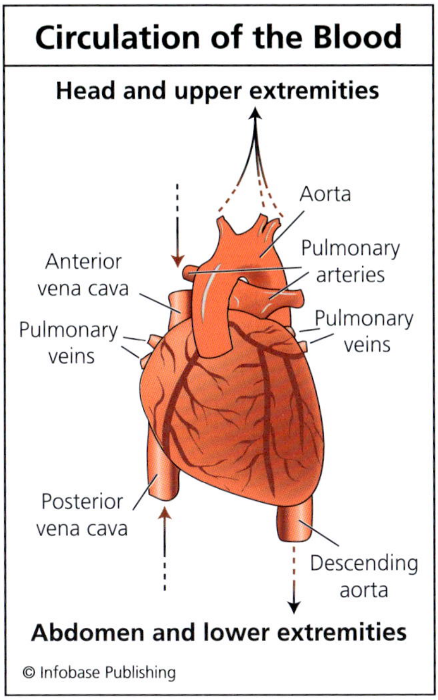

William Harvey was instrumental in understanding that Galen's theory about the blood was wrong.

that was produced—this process would require the body to create a vast quantity of blood at all times. As he reasoned his way through this, he concluded that the blood had to be recycling itself; the body could not be constantly producing new blood.

By 1616, he seemed to be further developing his theory though it took another 12 years before he published his findings in *Exercitatio Anatomica de Motu Cordis et Sanguinis in Animalibus* (An anatomical exercise on the motion of the heart and blood in animals), where he fully explained his belief that the blood was circulated by the heart within a closed circulatory system.

Unfortunately, Harvey's research notes were lost during the English Civil Wars. As a result, there are questions remaining about when Harvey knew what. Only his lecture notes from 1616 survive. While they provide an incomplete documentation of his work, at least they provide some insight into his process.

REACTION TO HARVEY'S THEORIES

Harvey's work attracted the attention of other physicians and scientists, but at first the reactions to him were very poor. He was attacked for taking issue with Galen, and no one felt his theory provided reason for any change in health care; bloodletting continued to be a popular treatment.

Harvey kept up with his research, pointing out that his evidence was observable and provable. He eventually began to gain a small following. One who came to believe in his theory was philosopher René Descartes (1596–1650) who was respected as one of the great scientific thinkers of the time and became one of Harvey's most prominent defenders. Descartes was younger than Harvey but working at about the same time. In addition, he wrote about anatomy and physiology. He was convinced that everything in nature could be described in terms of mathematics and science, and in 1647 he wrote *The Description of the Human Body* in which he suggested that the arteries and veins were pipes that carried nourishment around the body. It was not published until after Descartes's death in 1650.

From *New Discoveries at Jamestown: Site of the First Successful English Settlement in America* by John L. Cotter and J. Paul Hudson, 1957 *(U.S. Department of the Interior, National Park Service)*

Because Harvey was disheartened by the criticism of his work, he began to devote more time to practicing medicine and less time to research. As physician to James I and later Charles I, he had an exalted position from which he could work. Harvey accompanied the king on campaigns, took care of the royal family, and tended to the dying and wounded.

For 34 years, Harvey maintained his connection to St Bartholomew's where he developed a large private practice. As a physician he was very conservative in treatment and did not use many of the potent drugs of the time. For many years, he was one of the most trusted doctors in England, although publication of his theory on circulation in 1628 dealt a setback to his practice.

A REMAINING QUESTION ANSWERED BY MALPIGHI

The one question that Harvey could not resolve during his lifetime had to do with how the blood traveled from the arteries to

ON EMBRYOLOGY

In 1651, Harvey wrote a book that introduced his work in embryology, *De generatione animalium* (On the generation of animals), which was revolutionary for his time, but it did not attract the attention that his theories on circulation did.

Aristotle had taught that primitive organisms could reproduce via "spontaneous generation," and Harvey believed that all living things originated from an embryo that was found in the egg. He performed detailed examinations of chicken eggs at various stages. Once a hen laid a clutch of eggs, Harvey studied one egg per day, noting the changes that occurred from day to day. The earliest forms of life seemed to grow from a "scab" that was barely visible to the naked eye (and of course, he lacked the advantage of a microscope). He was not certain how the embryo was fertilized and with no way to magnify what he was studying, he never saw spermatozoa.

Following his study of chicken eggs, Harvey undertook a search for something comparable in mammals. He had come to believe that all animals must grow from a "spot of blood" that he called the "primordium." He felt the embryo developed its future parts slowly as it developed through what he called "*epigenesis.*"

Scientists of the period were certainly seeking answers to these questions, but the answer that took root for a long time was that of "preformation." This idea dated as far back as Plato, and it established that within each egg was a tinier egg and another miniature embryo within it that contained a even smaller egg with a smaller embryo—along the lines of the Russian nesting dolls.

Because there were no other good explanations, this idea, too, became established and was used to explain birth and creation until the late 18th century when Caspar Fredrich Wolff made progress in more fully establishing epigenesis as an explanation for the way an embryo grows.

the veins to return to the heart. That discovery, the discovery of the capillaries, was made by Marcello Malpighi (1628–94) of Bologna, using a very primitive form of the microscope. In addition to this major discovery, Malpighi founded the science of microscopic anatomy (he was the first *histologist*), which was to become an element of many fields of study, including physiology, embryology, and practical medicine.

Malpighi received his education at the University of Bologna and taught there before moving to Pisa and eventually going on to teach at other universities. Malpighi used an early microscope to study the skin and the kidneys, and he conducted the first species-to-species comparison of the liver. He was studying the lungs of a frog when he observed a network of tiny blood vessels—capillaries, minute vessels that link the end of the arteries with the beginning of the veins returning the blood to the heart. "I could clearly see that the blood flows through tortuous vessels," he wrote. His discovery of the capillaries was presented to the world in the form of two letters. *De pulmonibus* was published in 1661 and reprinted frequently after that. It provided the first account of the *vesicular* structure of the lung, and it made a theory of respiration possible.

Malpighi's Concept of Capillaries

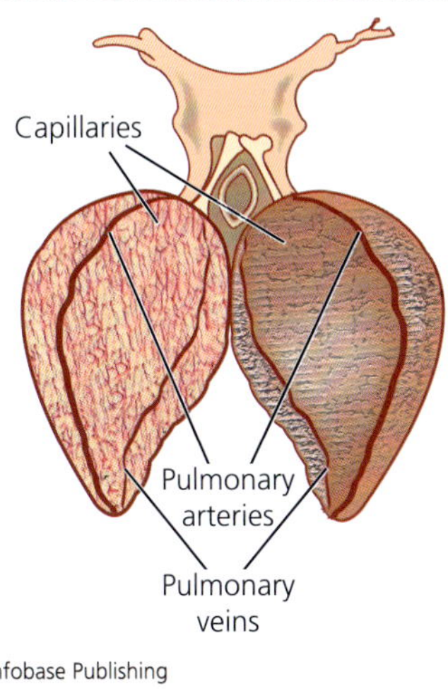

Without Malpighi's discovery of the capillaries, it would have been impossible to fully understand Harvey's theory of how the blood circulated.

His observations also led him to note the red blood cells; he was the first to do so, and he attributed the color of blood to these cells. It is indicative of the primitive state of the microscope that it took Malpighi another four years to reach a clear understanding of the *corpuscles* in the frog's blood. In 1666, his treatise *De polypo cordis* made an early effort to explain how blood clots and what clots are made of. Among his observations were the different clotting process in the right versus the left side of the heart.

As with so many others who broke new ground, Malpighi's discovery stirred up great controversy; others did not have the tools to verify what Malpighi saw, and they responded negatively from envy and lack of understanding. However, in 1668, his work attracted the attention of the Royal Society of London, and he began a correspondence with the society secretary that was eventually published. He went on to do detailed studies of the human tongue, noted the existence of taste buds, studied the anatomy of the brain, and saw the optic nerve. Some of the physiology of the digestive system was observed, including the bile secreted by the liver, and it was noted that the kidney functions as a filter. Malpighi also became fascinated with studying human fingerprints.

As Harvey had done, Malpighi studied the embryo and used a microscope to observe the development of the chick in its egg, verifying what Harvey had espoused. He later studied insects, particularly the silkworm, and noted that they do not use lungs to breathe; instead there are small holes in their skin called trachea. He also studied plants microscopically. In 1671, he published a book called *Anatomia Plantarum.* It was the most exhaustive study of botany at that time, although another work on botany by Nehemiah Grew (1641–1712) had been published just a few months earlier.

As he grew older, Malpighi's health declined, and in 1684, his villa burned and his microscopes and other scientific apparatus and his books and papers were destroyed. He was well respected, and Pope Innocent XII wanted to create a place for him. He was invited to Rome in 1691 to become a personal physician to the pope.

THE STUDY OF PHYSIOLOGY GROWS

Harvey lived long enough to see others build onto the experimental physiology that he inspired. Thomas Willis, Richard Lower,

Robert Boyle, Robert Hooke, John Mayow, and Christopher Wren strove for a better understanding of the human body.

- Thomas Willis (1621–75) was a prominent London physician who identified puerperal fever (childbirth fever) and began distinguishing among different forms of diabetes. Willis became known for his studies of the brain and diseases of the nervous system. He also identified what is now known as the circle of Willis, a system of connecting arteries at the base of the brain. He was among the group who helped found London's Royal Society in 1660 for the express purpose of furthering scientific study.
- Richard Lower (1631–91), a follower of Harvey and Willis, worked with Robert Hooke (see chapter 5) and published his findings regarding blood, including an understanding of the fact that the lungs were where the blood underwent change. It took another century before anyone discovered oxygen, so these fellows could not yet explain what was happening. Lower attributed it to *phlogiston,* a nonexistent chemical. He also experimented to find a way to transfuse from dog to dog and eventually from human to human (1665). At the time, it was believed that people could be helped by having old blood removed (bloodletting) and/or being infused with fresh blood. It was not easy to find people willing to undergo transfusions, but there were some. One eccentric scholar, Arthur Coga, agreed to have the procedure done before the Royal Society in 1667. Over time, transfusions began to be done somewhat more frequently, but in France they soon were overtaken by theological debates and the government of France prohibited transfusions. The work on blood transfusions could not move forward until Austro-American immunologist Karl Landsteiner (1868–1943) discovered the major blood groups and developed the ABO system of blood-typing.

■ Robert Boyle (1627–91) was a theologian, philosopher, and scientist who became best known for his work in physics and chemistry. Today, he is viewed as the first modern chemist. He recognized that the Greek definition of the elements (earth, air, water, fire) was incorrect, and he proposed a new definition of an element. In his work, he believed that experimenting was the key to learning, and he was devising a theory that everything was composed of minute but not indivisible particles of a single universal matter. Experiments that he conducted with his then-assistant Robert Hooke began to demonstrate that air was necessary for birds and animals to survive; he also noted that air caused iron to rust and copper to turn green. He and Robert Hooke (see chapter 5) are remembered for creating a vacuum pump that helped in the understanding of air.

Robert Boyle, engraving by George Vertue, 1684–1756; original artist Johann Kerseboom, d. 1708 *(Dibner Library of the History of Science and Technology)*

■ John Mayow (1641–79) worked in what is now sometimes called pneumatic chemistry, conducting early research into respiration and the nature of air. Drawing on Boyle's work on combustion, Mayow showed that fire derived strength from just part of the air, something he called "spiritus igneo-aereus." He determined that these particles were also consumed in respiration. He placed a lighted candle and a small animal in a closed vessel, and he noted that the candle went out and soon after the animal died. If there was no candle, the animal lived twice as long. Mayow concluded that there was something in the air that could be separated out by the lungs and passed into the blood. Essentially, Mayow preceded Priestley and Lavoisier by a century in recognizing the existence of oxygen.

- Christopher Wren (1632–1723) was an astronomer and mathematician who is primarily remembered for his work as an architect. He was largely responsible for designing the rebuilding of London after the Great Fire of 1666—he designed St Paul's Cathedral. His lectures as a professor of astronomy at Gresham College in London are actually where the roots of the Royal Society grew. Scientists gathered to hear his weekly lectures, and afterward they would talk. From this grew the ideas for the Royal Society for the "promotion of Natural Knowledge." His other endeavors involved optics, finding longitude at sea, cosmology, mechanics, microscopy, surveying, medicine, and meteorology. Wren also participated in the experiments with canine transfusions. In 1665, Wren was giving a dog an injection of opium using a bladder attached to a sharpened quill when he realized that injections could be given intravenously.

CONCLUSION

It took almost 50 years after the publication of Harvey's theory on circulation when teachers at the University of Padua introduced Harvey's ideas rather than Galen's, but from that time onward there was no turning back. The understanding of the circulatory system and Malpighi's early work with the microscope were of key importance in laying the groundwork for new fields of medical exploration.

5

The Microscope and Other Discoveries

Just as great progress was being made with inventions such as the printing press, the different ways to employ gunpowder, and the creation of the mariner's compass, scientists and inventors were also tinkering with ways to see things better through creating various devices that magnified objects. The development of the microscope was one of the most significant inventions of this period, and, over time, it was to have a major effect on medicine because suddenly scientists could see things they had never imagined.

Marcello Malpighi (see chapter 4) was among the first physicians to employ a microscope to good medical purpose. While William Harvey's theory about blood circulation (see chapter 4) made a great deal of sense, no one was able to explain how the blood went from the arteries to the veins until Malpighi used a microscope to discover the capillaries. But Malpighi's work offered only a peek at the possibilities, and a great deal more was accomplished by cloth merchant and hobbyist, Antoni van Leeuwenhoek, who was to greatly affect scientific discovery. His discoveries were verified and expanded upon by Robert Hooke, one of the leading

Leeuwenhoek and the "little animals" *(Department of Library Sciences, Christian Medical College—Vellore, History of Medicine Picture Collection)*

scientists of the day. As noted in chapter 4, Hooke investigated the contents of air under the guidance of scientist Robert Boyle and worked with Robert Willis on chemistry experiments. In addition, some of his most significant contributions were performed in his studies involving a microscope. The sight of small living things as viewed through magnifying devices renewed the debate about "where things come from." One of the popular theories of the time that will be examined in this chapter was that of "spontaneous generation."

In addition to these subjects, this chapter will set the scene for what was happening with several diseases that were becoming increasingly common between 1450 and 1700 as an increase in sea voyages resulted in greatly expanded exposure to the world. *Scurvy* was an increasing problem, particularly among sea travelers, and smallpox, which had been around for a time, was becoming more prevalent and more deadly.

THE DEVELOPMENT OF THE MICROSCOPE

Simple magnification works by viewing something through a *convex* lens that is thicker in the middle, rounding out toward the edges. Understanding the nature of magnification almost certainly came about by happenstance. Someone probably picked up a piece of transparent crystal that happened to be convex and must have noted that things appeared larger than they were if viewed through the crystal. As early as the Roman era, they wrote of "burning glasses," so they had learned that by focusing the Sun's rays through a translucent substance, one could set fire to a piece of parchment or cloth.

The first documented use of a convex lens for magnifying images appears in the *Book of Optics* written in 1021 by Abu

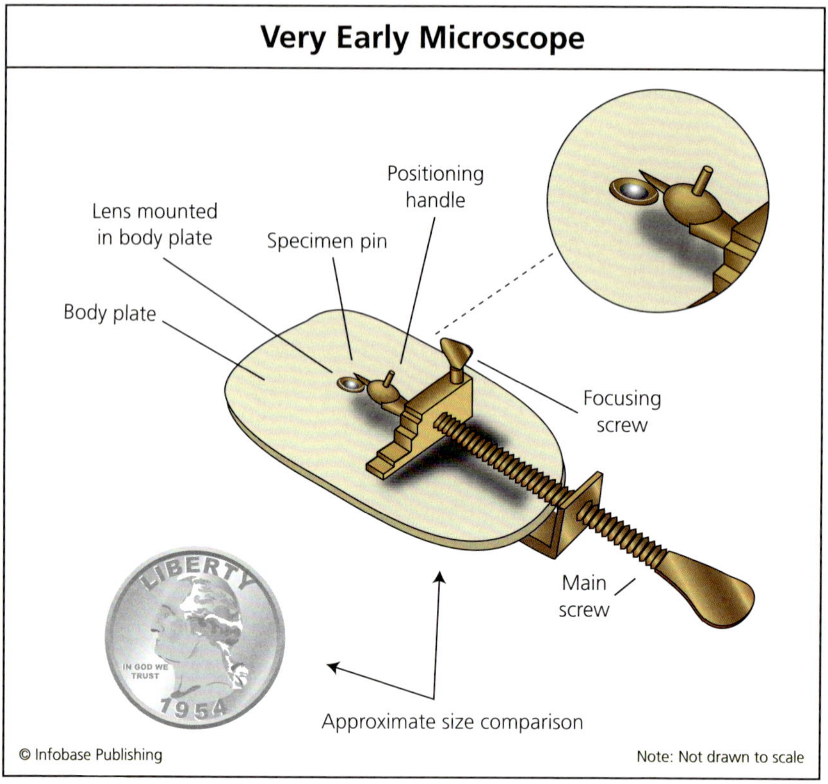

Early microscopes took many forms. This is an example of one of them.

Ali al-Hasan Ibn al-Haytham (965–1039). During the 12th century, English scientist Robert Bacon (1220–92) described using a magnifying lens; scientists were learning that the rate of magnification could be altered by adjusting the placement of the glass in relation to the object being viewed. By the 13th century, Italians had begun to create lenses that could be worn as eyeglasses.

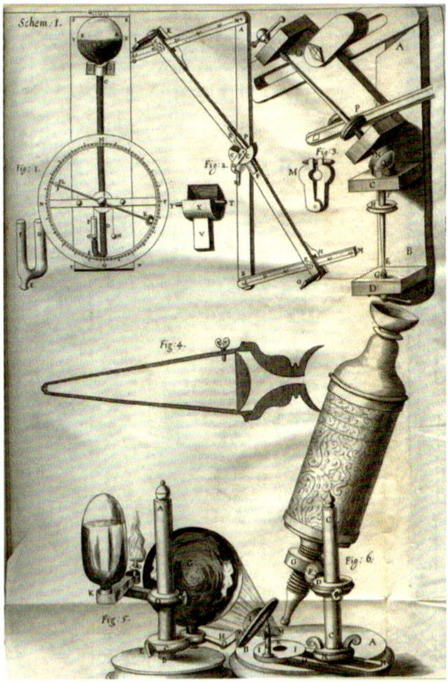

From *Micrographia: Or Some Physiological Descriptions of Minute Bodies Made by Magnifying Glasses* by Robert Hooke (1635–1703) *(Stanford School of Medicine)*

The next major advance in the field occurred in 1590, when two Dutch spectacle makers, Zaccharias Janssen and his son Hans, were experimenting with using different lenses within a tube. They saw that by combining the lenses in a particular manner, nearby objects could be magnified to a higher degree than what could be accomplished with a single lens. This was the beginning of the compound microscope. Zaccharias Janssen also developed a method for long-distance viewing by using a longer, bigger tube and arranging the lenses differently to create what became known as a telescope. In 1609, Galileo adapted the Janssens's creation to design an instrument that also had a focusing device.

These early microscopes magnified objects approximately 10 times their original size. They were called flea glasses because by looking through them one gained a much better view of very small insects. However, these devices were still so primitive that a well-ground magnifying lens was more effective. Despite efforts

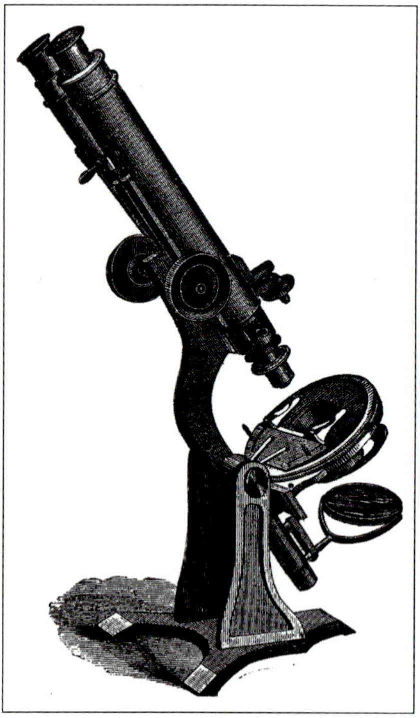

Early microscope *(Project Gutenberg)*

by Galileo, Robert Hooke, and Jan Swammerdam (from the Netherlands), the highest level of enlargement with these devices was a magnification of 20 to 30 times.

Antoni van Leeuwenhoek, the Dutch cloth merchant, used simple magnifying lenses in his commercial work to ascertain the thread count in cloth. He became fascinated with the process and created new ways to grind and polish very small lenses and give them additional curvature, all of which increased the level of magnification. Leeuwenhoek was also very sensitive to capturing the light when using the lenses, and this greatly enhanced what he could see. He was able to achieve a magnification that enlarged things 200 times. This explains why Malpighi, working with a microscope, could see capillaries, while Leeuwenhoek, using finely ground but powerful magnifying glasses, could see "little animalcules."

Today the basic microscope is still of value as there are always objects that don't necessitate "super power" magnification. The device used today is very similar to the design of the microscope that existed in the 19th century. Today, scientists can also benefit from seeing what would have been unthinkable in Leeuwenhoek's day by using the electron microscope that uses beams of electrons focused by magnetic lenses instead of rays of light, the magnified image being formed on a fluorescent screen or recorded on a photographic plate: Its magnification is substantially greater than that of any optical microscope.

LEEUWENHOEK AND HIS LENSES

Antoni van Leeuwenhoek (1632–1723) was an unlikely fellow to move science forward in such a major way. Leeuwenhoek was a tradesman born into a family of tradesmen who lived in Delft, Holland, where he worked as a fabric merchant. He was intensely curious about many things, and, using one of the tools of his trade—the magnifying glass that merchants used for checking thread count—he began the serious pursuit of a hobby. He was fascinated by what was beyond the vision of the naked eye. He ground his own lenses, and in 1674 he made one of his first significant discoveries when he was able to see red blood corpuscles. Leeuwenhoek had acute eyesight and understood how to direct light onto an object. Key to advancing his discoveries was Leeuwenhoek's ceaseless fascination with creating better and better lenses. Leeuwenhoek created some 500 different lenses, eventually finding one that enlarged items about 200 times their natural size, which led him to make one that was powerful enough to see microorganisms. In 1677, he spied never-before-seen spermatozoa, and in 1683 he provided an accurate description of the capillaries. Leeuwenhoek studied animal and plant tissues as well as mineral crystals and fossils; he was the first to see microscopic animals such as nematodes (roundworms) and

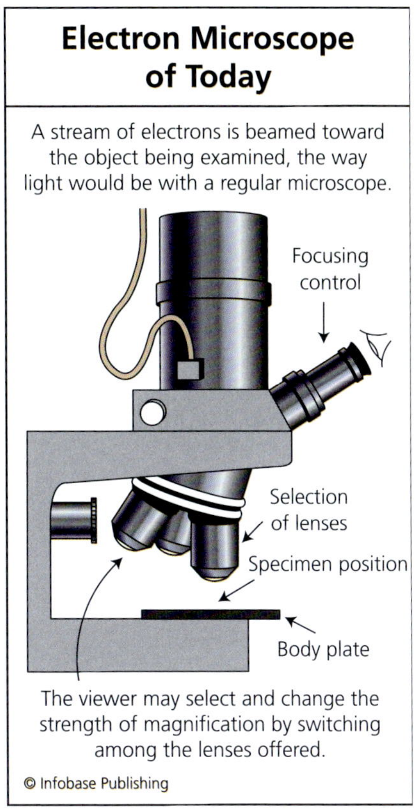

An electron microscope uses electrons to illuminate a specimen and to create an enlarged image, but the electron microscope was not invented until the 1930s.

rotifers (multicelled animals that have a disk at one end with circles of strong cilia that often look like spinning wheels) as well as blood cells and sperm. Leeuwenhoek worked meticulously, and wanting to accurately document what he saw, he hired and worked with an artist to illustrate his findings.

A well-respected physician in Delft, Regnier de Graaf, aware of Leeuwenhoek's work, realized the significance of it. In 1673, he wrote a letter to Henry Oldenburg, secretary of the Royal Society (see chapter 4) in London, about what the cloth merchant was doing. Oldenburg wrote directly to Leeuwenhoek, requesting further information. Leeuwenhoek's first letter, dated April 28, 1673, detailed his microscopic observations of mold and bees. This was the beginning of a correspondence with the Royal Society that was to consist of more than 165 letters, written until the end of Leeuwenhoek's life.

The Little Animalcules in Tooth Plaque

Ten years after the Leeuwenhoek-Royal Society correspondence began, Leeuwenhoek wrote to the Royal Society about his observations of the content of tooth plaque. (In those days, tooth brushing would not have occurred regularly—if at all—as no one would have understood the advisability of clean teeth.) Leeuwenhoek's observations of living bacteria were the first ever recorded. In 1683 Leeuwenhoek wrote

> I then most always saw, with great wonder, that in the said matter there were many very little living animalcules, very prettily a-moving. The biggest sort . . . had a very strong and swift motion, and shot through the water (or spittle) like a pike does through the water. The second sort . . . oft-times spun round like a top . . . and these were far more in number . . .

In the mouth of one of the old men whose plaque he studied, Leeuwenhoek found "an unbelievably great company of living animalcules, a-swimming more nimbly than any I had ever seen up to this time. The biggest sort . . . bent their body into curves in going forwards . . . Moreover, the other animalcules were in such enormous numbers, that all the water . . . seemed to be alive."

Leeuwenhoek wrote all his letters in his native Dutch, and he never published the information in book form. However, scientists—after having Leeuwenhoek's work investigated by Robert Hooke—soon realized that what this tradesman was doing was quite remarkable. They had his descriptions translated from Dutch into English or Latin, and his findings were regularly published in the Royal Society's publications. In 1680, they made him a member of the Royal Society. He became famous all over Europe and in 1698 was asked to demonstrate circulation of the blood in an eel before Peter the Great of Russia.

In a letter dated June 12, 1716, Leeuwenhoek explained the reasoning behind his work: "My work, which I've done for a long time was not pursued in order to gain the praise I now enjoy, but chiefly from a craving after knowledge, which I notice resides in me more than in most other men. And therewithal, whenever I found out anything remarkable, I have thought it my duty to put down my discovery on paper, so that all ingenious people might be informed thereof."

None of the lenses he created have ever surfaced. Perhaps because Leeuwenhoek used gold and silver to make his instruments, his family sold them after he died.

ROBERT HOOKE: FORGOTTEN GENIUS

Robert Hooke (1635–1703) delved into so many fields (physics, astronomy, chemistry, biology, geology, architecture, and naval technology) in which he made major contributions that some consider him the single greatest experimental scientist of the 17th century. His level of genius is sometimes compared to that of 14th-century Leonardo da Vinci. Among his major contributions was to create the anchor escapement and balance spring for use in timepieces, both of which increased the accuracy of clocks. He also devised a theory of elasticity that is still used today and wrote a remarkable book describing what he saw through the microscope.

LIVING THINGS FROM NOWHERE

Now that bacteria had become visible, scientists needed to continue to wrestle with the very logical question: "Where do bacteria come from?" One explanation of how matter "appeared" had been discussed long before Leeuwenhoek. Both the Greeks and the Egyptians had believed that some living things could arise from nothing . . . what was called "spontaneous generation" or "*abiogenesis.*" A good example came from Egypt. When the Nile River flooded each spring, nutrient-rich mud covered the river banks, and soon the fertile land along the water's edge was filled with frogs. The Egyptians concluded that mud gave rise to frogs. In medieval Europe, farmers followed similar thinking when it came to big increases in the mouse population. Farmers stored grain in barns with thatched roofs that often leaked, and the grain became moldy. Mice hovered around any grain-filled areas, so the belief arose that mice actually came from moldy grain.

As Leeuwenhoek began writing of the tiny, rapidly moving "animalcules" he spied in miscellaneous sources ranging from rainwater, liquid in which he had soaked peppercorns, and scrapings from teeth, his findings seemed to verify this theory of something living coming out of nothing.

To pursue this line of thinking, Italian physician and naturalist Francesco Redi (1626–97) designed a series of experiments to see if he could prove this theory one way or the other. Working from the premise that many believed that

Hooke's Life

Hooke was born in 1635 at Freshwater, on the Isle of Wight, the son of a churchman who largely provided his early education. He eventually went on to Oxford where he encountered some of the maggots were generated from rotting meat, Redi filled six jars with decaying meat; three were left open and three were tightly sealed. The unsealed jars soon attracted flies that laid eggs on the meat; the sealed jars were impenetrable so no flies crawled on the meat. Soon, maggots developed on the meat in the open jars, and Redi proclaimed that he had proven that spontaneous generation could not occur. Believers of the theory still did not agree. They claimed that the lack of air on the sealed meat was all that kept the spontaneous generation from occurring.

To counter this argument Redi repeated the experiment, but he used a tightly woven net over the sealed jars instead of something impenetrable. This permitted air to reach the meat, but not the flies. Even this did not convince those who wanted to believe otherwise. As they saw it, "little animacules" like what Leeuwenhoek had spotted were very simple creatures that could be generated from nonliving material.

The debate over spontaneous generation raged on with scientists taking both sides and devising their own experiments to prove their points. The debate did not end until the 19th century when the Paris Academy of Sciences offered a prize for the scientists who could bring resolution to what had become a very contentious issue. In 1864, Louis Pasteur was recognized for proving that living things could not generate "out of nothing."

best scientists working at the time. Well-regarded chemist Robert Boyle took him on as an assistant (from 1655–62) and they worked together on the creation of the vacuum pumps that let Boyle explore the composition of air. While Boyle did not succeed at identifying oxygen, the work they performed opened the door for later discovery. While working with Boyle, Hooke also demonstrated that a dog could be kept alive with its *thorax* opened, provided that air was pumped in and out of its lungs.

He soon attracted the attention of other scientists for his skill at designing experiments and building equipment for use during the testing phase. In 1662, Boyle released Hooke from his duties, and Hooke was given a staff position at the newly formed Royal Society as curator of experiments. This job involved translating the ideas developed by members of the group into experiments that could test the scientists' theories. At the weekly meetings, these experiments were then demonstrated by Hooke so that they could be observed by all in attendance and discussed. Later, Hooke became professor of geometry at Gresham College in London where he continued to pursue his many interests, but he was still the person the Royal Society members turned to for carrying out their experiments. (He performed these responsibilities for 40 years, first from a staff position, and later as a fellow.)

Like other notables of his day, Hooke worked in more than one profession, and like fellow member Christopher Wren, Hooke was an architect. When the Great Fire of London devastated the city, Hooke worked as chief surveyor to help rebuild the city.

HOOKE'S WORK IN MICROSCOPIC MATTERS

Though Leeuwenhoek is generally referred to as the father of microscopy, Hooke, too, is often given this mantel. Hooke created a compound microscope and illumination system that was one of the best of the time, and it was used to demonstrate findings at the Royal Society's meetings. He used it to observe insects, sponges, *bryozoans, foraminifera,* and bird feathers.

Hooke was the first to coin the term *cell* to describe the basic unit of life. In 1665, he cut a sliver of cork through a microscope lens and noticed "pores." Hooke concluded that the "pores" had served as containers for the "noble juices" or "fibrous threads" of the once-living cork tree. This was the first discovery of plant cells, and Hooke used the term *cell* because the boxlike cells of cork reminded him of the cells of a monastery. When describing the thin slices of cork he examined, he noted: "I could exceedingly plainly perceive it to be all perforated and porous, much like a Honey-comb but that the pores of it were not regular . . . these pores, or cells, . . . were indeed the first microscopical pores I ever saw, and perhaps, that were ever seen, for I had not met with any Writer or Person that had made any mention of them before this . . ." While this was a revolutionary discovery, Hooke also reported seeing similar structures in wood and other plants, but he felt that cell structure was limited to the structures in plant material.

Hooke published *Micrographia* in 1665 with detailed illustrations and complete and accurate descriptions of his observations using the microscope. To make the book accessible to as many people as possible, Hooke wrote it in English not Latin. The book was a best seller, although many made fun of him for paying attention to finding "mites in cheese." Others, like Samuel Pepys, a government official and diarist, called it the "most ingenious book that I have ever read in my life."

Hooke was vital to Leeuwenhoek's fast rise through the world of science. In 1678, when Leeuwenhoek corresponded with the Royal Society about his "little animalcules," the society turned to their trusted scientist Robert Hooke to investigate Leeuwenhoek's findings. Hooke verified the "little animalcules," the bacteria and protozoa that Leeuwenhoek claimed to have seen, which certainly put Leeuwenhoek's findings on a much faster track to acceptance than would otherwise have occurred. While Hooke remarked on the clarity

How Cells Were Viewed in the 1700s

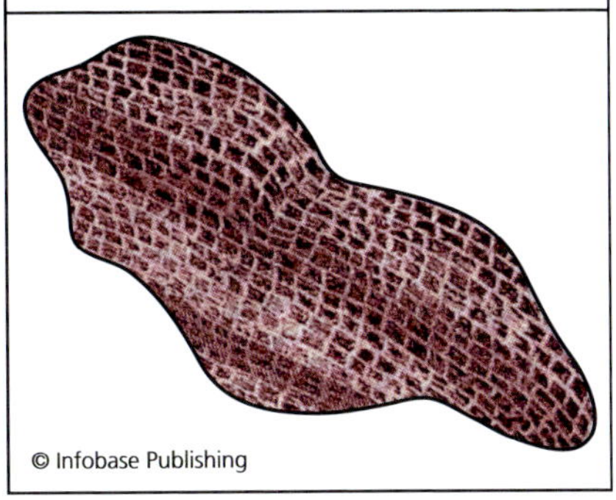

The magnifying power of early microscopes was not very strong, and this would have been how cork might have looked.

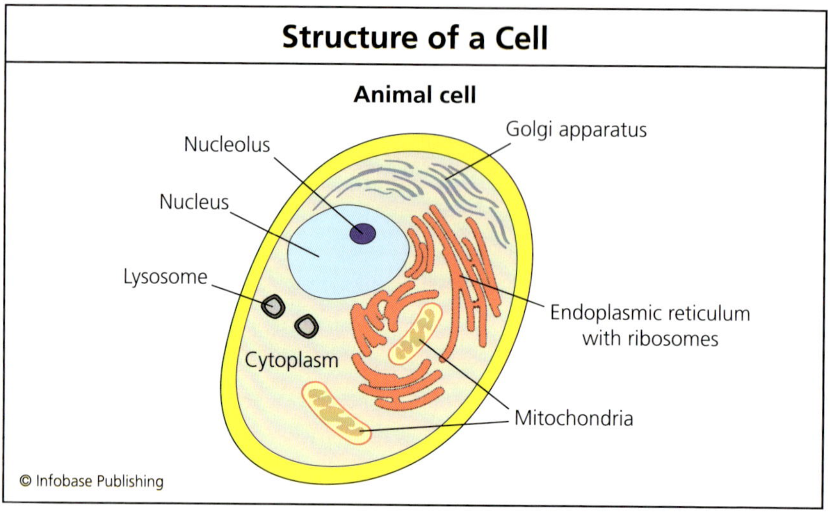

Today, scientists can magnify to the point where they can determine specific parts of a cell.

of Leeuwenhoek's simple microscopes and noted that they were actually superior to what he could see through his compound microscope, he noted that he personally found them much more difficult to use.

Hooke also used his microscopes to study fossils and geology. During the 17th century, there was no understanding of what a fossil was. Since the time of Aristotle, it had been believed that fossils somehow formed and grew within the earth. Even well-respected naturalist and shell collector Martin Lister, a contemporary of Robert Hooke, felt that fossils were simply a type of stone. Hooke used his microscope to examine various fossils and noted that there were striking similarities between things like fossilized shells and recently found mollusk shells. He noted that the shell-like fossils he examined were "the Shells of certain Shel-fishes, which, either by some Deluge, Inundation, earthquake, or some such other means, came to be thrown to that place and there to be fill'd with some kind of Mud or Clay, or petrifying Water . . ."

Hooke theorized that living things could be turned into stone (fossils) by mineral-rich water washing over them, leaving behind mineral deposits over a long period of time. Two and a half centuries before Darwin, he concluded that fossils are not accidents of nature but the remains of once-living organisms, and they provide a traceable record of how organisms have transformed over time.

Today, Hooke is acknowledged as one of the preeminent scientists of the 17th century, but shortly after his lifetime he was all but forgotten. Isaac Newton had become president of the Royal Society, and Newton vehemently disagreed with Hooke's demonstrations for the society on gravitation. Though the true story of what happened is not really known, some speculate that Newton did what he could to obscure the work of the other scientist.

One frustration of modern historians is the fact that there is no portrait or depiction of Hooke. Though there had been one rendering of him in a stained-glass window at St Helen's Church, it was destroyed in a 1993 bombing of the Bishopgate area by the IRA. His microscope, however, a leather and gold-tooled one made by Christopher White in London, is on display at the National Museum of Health and Medicine in Washington, D.C.

THE RISE OF SCURVY

Scurvy was an illness that had been around for a long time. The disease frequently presents with spots on the skin (mostly on the legs) and features bleeding from the mucous membranes and spongy gums, resulting in tooth loss. As the disease progresses, a victim's muscles become rubbery, making it hard to move around. It was recognized and described as early as Hippocrates, and it was widely reported in the 13th century among those who joined the Crusades.

In the 15th and 16th centuries, scurvy appeared again in a prominent way because of the increase in sea travel. Today, it is known that scurvy results from a deficiency of vitamin C, a vitamin that is easily obtained from fresh fruits and vegetables. Sailors embarking on voyages that might take one to two years could not carry much fresh food with them, so vitamin C deficiencies and scurvy abounded.

Jacques Cartier (1491–1557), a French explorer and discoverer of the St. Lawrence River, encountered circumstances that permitted him to investigate how to prevent or cure scurvy. Cartier was on what was the second of three voyages he made to Canada on behalf of the French king. It was the autumn of 1535, and he and his men had explored the St Lawrence, going as far south as what is now Montreal. Ignoring warnings from the Native people about the risk of the river freezing, Cartier delayed his departure to the ocean to return to France. Near what is now Québec, the French ship was halted by ice, and they were destined to spend their first winter in the New World. The Frenchmen established a very basic fort near a village of Iroquois, but as the days wore on the Indians began getting sick from what seemed to be diseases carried by the Frenchmen. The natives responded to this threat by trying to cut off contact.

Both groups had little access to anything fresh to eat, and over time, many of the crew—and Cartier observed some of the Indians—came down with an illness that featured open sores, bleeding gums, and muscle weakness. However, Cartier also saw the chief's son, Dom Agaya, was out and about and seemed to be doing fine. Cartier finally approached the chief's son and asked for help. Reluctantly, the chief finally agreed to allow his son to show Cartier some of the Iroquois' secret medicines. Dom Agaya then demonstrated for Cartier how he stripped leaves from a white cedar tree and boiled the leaves to make a tea. Upon being presented with a cup of the tea, Cartier refused it, remembering the Iroquois' animosity and worrying that they would try to poison him. To some of the crew who were desperately ill, a quick death from poison seemed like an acceptable alternative to a long slow but certain death from scurvy. As it happened, the men who drank the tea felt better quite quickly. (Today the remedy has been analyzed, and the leaves from the eastern white cedar can be brewed to provide 50 mg of vitamin C per 100 grams.) Cartier used the remedy for all of his men, and they overcame their scurvy. In return, he did what he could to help the Native people with their illnesses. (Sadly, Cartier's behavior did not continue to be exemplary. When he wanted the chief to return with him to France to tell the king about the riches of the area, the chief refused, and Cartier and his men captured the chief and his sons and took them by force. The chief never made it back to his homeland.) While the remedy provided Cartier was effective, it was not yet understood why.

By 1614, British physicians were beginning to understand that scurvy resulted from a dietary deficiency and that consuming citrus fruit could improve the condition. However, they felt the key was the acid component, so when ascorbic acid (vitamin C) was not available, they felt that oil of vitriol (sulfuric acid), which is actually a highly corrosive chemical that today is used in products like fertilizer, was an acceptable substitute.

SMALLPOX TAKES ON NEW VIRULENCE

In 16th-century Europe, smallpox was common but rarely fatal, and in many communities it was considered to be a part of the types of illnesses suffered in childhood; usually with the young suffering milder cases. During the 17th century, smallpox became more virulent, with some scientists speculating that a new strain may have appeared from Asia. Various areas were devastated by epidemics. Italy was particularly hard hit over a 20-year period in the 16th century (1567, 1570, 1577, and 1588), and in England there was a particularly bad epidemic in 1659. Physicians had few ideas about prevention and disagreed on what to do, and nothing worked reliably. The disease mainly followed its own course, with a very few getting better, but most either dying or being maimed or blinded by the illness.

The ability to travel farther than ever before meant that ships carried to other lands contagious diseases to which Native populations had little or no *immunity.* The effects were often devastating. In Central America, almost half the Aztec population perished after the arrival of Spanish conquerors in what would be the first smallpox epidemic in the New World. In 1531, another ship came, and the new soldiers exposed even more people to the disease. In 1576, the Aztec population fell from 25 million (in 1492) to an astonishing 2 million. Similarly, the Inca population of Peru fell from 7 million to 1.5 million. It is debatable whether *variola major* (smallpox) arrived on its own or whether *variola minor* (measles), carried by an asymptomatic Spanish soldier, just turned virulent in the new population.

One of the earliest medical documents printed in America north of Mexico was Thomas Thacher's broadside *A brief rule to guide the common-people of New-England how to order themselves and theirs in the small-pocks or measles* (1677–78).

When disease brought devastation, physicians and scientists were inspired to develop treatments as well as methods that might prevent people from becoming sick. Paracelsus (see chapter 1) was among those who traveled widely and learned from other cultures, and, while visiting Constantinople in 1522, he learned of a peasant remedy to prevent smallpox. The people scraped off the smallpox scabs of someone who had smallpox and created a powder that could be inhaled or injected. By introducing a small amount of the disease, it sometimes helped ward off a more serious case of it when the illness spread throughout a community. Paracelsus's method did not become accepted, possibly because he also used this method—with no success—with other illnesses.

Portrait of Mary Wortley Montagu by Charles Jervas *(National Gallery of Ireland and the Yorck Project)*

Another person intent on prevention in order to protect her children was Lady Mary Wortley Montagu (1689–1762), the wife of the British diplomat to Turkey (see chapter 1). Montagu had observed Turkish women hold "smallpox parties" where the children were met by an old woman with the "nutshell full of the matter of the best sort of smallpox and asks what veins you please to have open'd. She immediately rips open . . . and puts into the vein as much [smallpox] matter as can lie upon the head of her needle." As their bodies fought off the small dose of it, they were able to resist becoming ill when exposed to the illness again. Lady Montagu had her own daughter *inoculated* with this method (1721), and, when she returned to England, Lady Montagu recommended that doctors use the method, but it did not really catch on.

CONCLUSION

The ability to magnify through the use of magnifying lenses and the invention of the microscope were to change the study of science and medicine forever. Suddenly small things—from the capillaries in the body to the Leeuwenhoek's "little animalcules" could be seen and studied in a way that had never been possible before. Robert Hooke's deduction that what he was seeing under the microscope represented the "building blocks" of living things was another big step forward.

Scurvy and smallpox were not new to this era, but they made fresh appearances that tested those who wanted to develop treatments. Though no certain cures were found, the groundwork was laid for additional progress that was to be more fully realized in the next century.

6

Syphilis and What It Reveals of the Day

The changing pattern of illness following the Middle Ages became almost as significant a marker in the shifting of an era as was the transformation of art and science. Starting in the 15th century, Europeans began making long voyages and extending their boundaries through commercial expansion and warfare. In the process, previously isolated people were suddenly immersed in a broadening germ pool. Though the incidence of leprosy, so prevalent during the Middle Ages, began to fade, syphilis, typhus, smallpox, and influenza became major threats to public health.

Syphilis was a real scourge for the population. While venereal diseases had been around since antiquity, it is currently believed that syphilis did not exist in Europe until it was first diagnosed in 1495. The army of King Charles VIII of France launched an attack on Naples, Italy, and the French soldiers started falling ill so quickly that Charles had to halt the fighting and abandon his attempt to conquer Italy. The soldiers, who are thought to have brought the illness from the New World, infected those in the area where the attack was underway; those who survived until after the battle carried the disease throughout Europe as they returned home. Because it began within the French army, syphilis became known as *morbus Gallicus* (the French disease or the French pox). The French did not like it being called the French disease, so they called it the Neapolitan disease. In Germany, it was referred to as *die Blattern.*

This chapter will take a look at how syphilis may have begun, how it spread, and how it adapted in order to continue its existence. The effect it had on cities and towns as they looked for ways to control it will also be discussed.

SYPHILIS

Syphilis is now known to be caused by a spirochetal bacterium *Treponema pallidum.* It is similar to four clinically distinct human diseases, and some bacteriologists feel that the *spirochete* mutated over time. The other forms were not sexually transmitted and tended to occur in children who lived in warm climates. As people migrated to temperate areas and needed clothes, fewer children acquired skin ailments that seemed to pass via skin-to-skin contact. As a result, more people reached adulthood without acquiring immunity, and the disease may have mutated into a sexually transmitted disease.

One of the problems with syphilis was that it was difficult to diagnose without the advantages of modern science. Because the lesions can look like leprosy, tuberculosis, scabies, a fungal infection, and various skin cancers, it became known as the great mimic. Untreated, syphilis goes through the following three stages:

1. It begins with a small lesion (*chancre*) on the part of the body where the infection first appears (often the genitals), which may ulcerate or disappear.
2. The next stage may arrive within weeks or months of the first infection. It is more systemic, involving fever, headache, sore throat, localized rash, skin lesions, swollen lymph nodes, mouth sores, and bloodshot eyes. It may go away with little or no treatment.

John Misabaun and Richard Rock argue over treatment, while Moll Hackabout dies of venereal disease.

3. The third stage involves a chronic obstruction of small blood vessels, abscesses, and inflammation that may permanently damage the cardiovascular system and other organs. Syphilis may affect neurological function, causing impaired vision, loss of muscular coordination, paralysis, and insanity. Women frequently have trouble conceiving, and they may suffer miscarriages, stillbirths, or give birth to a deformed child.

There is no predictable pattern for the various stages; each case is different, making it all the more difficult to diagnose and treat.

The Spread of Syphilis

Scientists who have studied the evolution of syphilis note that syphilis began as an acute, debilitating disease that would have repelled potential sexual partners because the victims became so sick. Despite this, syphilis started out as a disease that spread quickly. It reached England in 1496, Poland in 1499, Russia in 1500, and China by 1505. It is thought that it was carried to India by one of Columbus's men who traveled with Vasco da Gama to explore a new route eastward. The disease took longer to reach more remote areas. Japan finally saw its first case in 1569, and Iceland held off until 1753. It also spread virulently within communities. By the end of the 16th century, one-third of all Parisians were infected with syphilis.

Within 50 years, it evolved into a milder form. Dr. Robert Knell at Queen Mary's School of Biological Sciences at the University of London notes that when syphilis first appeared it was too virulent for its own good. Many of the early symptoms of the epidemic—such as disfiguring pustules on the face accompanied by a foul smell—would have been obvious to any potential sexual partners of a sufferer, enabling people to avoid the infected person and reduce transmission. Other symptoms, such as agonizing pains in the joints, would have effectively disabled the sufferer, distracting them from seeking out new sexual partners. As a result, less virulent strains of the disease were transmitted more often, thus leading to changes in the severity of the disease.

"Syphilis changed from a virulent disease to a relatively mild one in a very short period," says Dr. Knell. In diseases that course through animal populations, a change in virulence can occur, but this is the first time such a dramatic change has been documented in a human disease. Knell notes that understanding this type of shift in a disease could be vital in understanding how dangerous new diseases such as SARS (severe acute respiratory syndrome) or Ebola might change if they become endemic in the human population.

THE POSSIBLE ORIGINS OF SYPHILIS

There have been many theories about how syphilis traveled to Europe. (While the first documented evidence of the disease appears in the 1490s, scientists are always mindful that they must try to identify whether there is evidence of syphilis in some form in Europe before this time.) The earliest explanations of syphilis came from medical astrologers. Because magic and astrology were still important components of disease, this was in line with the thinking of the day. Astrologers suggested that it was caused by a

HOW THE DISEASE CAME TO BE CALLED SYPHILIS

Girolamo Fracastoro (1478–1553) was an Italian physician, astronomer, and poet. (Fracastoro is discussed in greater detail later in the chapter.) He maintained a private medical practice in addition to teaching, and he was particularly fascinated by epidemic diseases, which became the focus of some of his studies. In 1530, he published a poem *Syphilis sive morbus Gallicus,* which described a disease suffered from by a handsome young shepherd Syphilis. As Fracastoro told it, Syphilis brought about the illness because he cursed the Sun. To punish men for this blasphemy, Apollo, god of the Sun, shot deadly rays of disease at the Earth. Syphilis was the first victim but it soon spread to others, including the king. The story soon became very popular and went through many editions.

Within the Latin verse, Fracastoro presented the symptoms, the pattern of the disease, and the recommended treatment. As a result, the signs of the illness became better known, and eventually the shepherd's name was adopted as the name of the disease.

Fracastoro also pointed the way to one of the early cures. Another verse described a gardener who was ill being ushered to a cavern where he was bathed in a river of *quicksilver* (mercury). Fracastoro himself mixed mercury with black hellebore and sulfur to use on patients. The patient's body was entirely covered with the mixture, and then he was wrapped in wool and kept there until the disease was flushed from the body through sweating.

malign conjunction of Jupiter, Saturn, and Mars that occurred in 1485. This produced a poison that spread throughout the universe, which caused the illness to spread.

Others felt that syphilis was a "venereal leprosy," a combination of leprosy and gonorrhea, but the evidence here is ambiguous. Another theory holds that syphilis was brought to Portugal by slaves who were brought to Europe from Africa after Prince Henry sent expeditions to explore the coast of western Africa in 1442. There is complete documentation of an African disease called yaws that features skin lesions similar to syphilis. Though yaws is often compared to syphilis, yaws begins with skin lesions, as syphilis does, but does not follow the same disease pattern. Scientists must fully examine whether yaws somehow mutated from a disease that spreads skin-to-skin, primarily among children in warm climates, or whether syphilis was something entirely different.

Another popular circumstantial theory is that Columbus and his men brought back syphilis from the New World. If syphilis did not exist in Europe in any form before this date, then the timing of this incident adds credibility. Evidence exists of a similar disease which Natives in the West Indies suffered from. Columbus traveled to what he continued to think was the Indies in 1492, and four months later he and some of his crew returned to Spain with ten Natives, one of whom died shortly after arrival. A contemporary physician adds credence to this belief. Rodrigo Ruiz Díaz de Isla (1462–1542) was a Spanish physician who said that he treated several sailors in 1493 who suffered from a strange disease involving skin eruptions. Later, the Spanish historian and writer Gonzalo Fernández de Oviedo y Valdés noted that some of the sailors who had been in the New World had gone home to join the French army for their attack on Italy, where the disease was first noted, adding another fact that fits with the circumstantial evidence.

In 2008, researchers at Emory University undertook newly possible genetic studies—phylogenetics, the study of evolutionary relatedness between organisms—to examine various geographically different strains of disease. The venereal strain of syphilis that spread in Europe seems most closely linked to a strain of yaws

Gerard de Lairesse by Rembrandt. Lairesse suffered from congenital syphilis, and his swollen features and bulbous nose are signs of the disease. *(Metropolitan Museum of Art and the Yorck Project)*

that existed in South America. This would support the hypothesis that syphilis or a related illness originated in the New World.

Though not all researchers agree, one of the coauthors of the Emory study, skeletal biologist George Armelagos, says: "Syphilis was a major killer in Europe during the Renaissance. Understanding its evolution is important not just for biology, but for understanding social and political history." He goes on to note its relevance today: "It could be argued that syphilis is one of the important early examples of globalization and disease, and globalization remains an important factor in emerging disease."

TREATMENT THEORIES

Over time, physicians began to use mercury as a cure (see *How the Disease Came to Be Called Syphilis*), combining it with other ingredients including lard, turpentine, incense, lead, and sulfur. One physician Giovanni de Vigo (1450–1525) decided that live frogs were a good addition though it is not clear exactly how the frogs were used. Those with syphilis sat in a tub in a hot, closed room where they could be rubbed with mercury ointments several times a day. Shakespeare notes the torments of syphilis and makes reference to the "tub of infamy." (The nursery rhyme "Rub-a-Dub-Dub" is thought to be about syphilis.) As a result, mercury became strongly associated with the illness and was used until the 1940s. However, few physicians left it at mercury. They added purgatives and tonics and provided bizarre dietary restrictions.

Today, it is known that mercury is actually quite toxic, but this was not known at the time. Though Bernardino Ramazzini (1633–1714) wrote *On the Diseases of Workers* and noted that mercury seemed to bring about ill effects, it was not until the 19th century that they realized that excessive salivation and mouth ulcers were signs of mercury "irritation," not the sign of someone recovering from syphilis.

Another treatment that came from the New World was guaiac, also known as holy wood. The wood came from evergreen trees that were *indigenous* to South America and the West Indies. Those who used it felt that if the disease came from the New World then so should the treatment. It soon developed that the rich used holy wood and the poor used mercury.

Today, venereal disease victims are reluctant to discuss their ailments, but this was not the case in the 16th century. At that

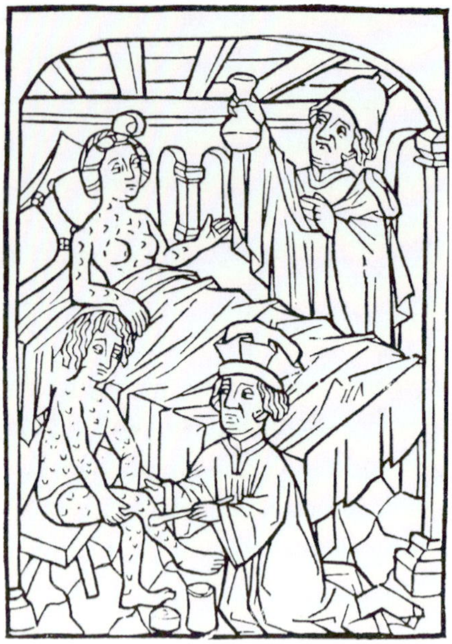

Application of mercury (for syphilis), after a painting by Bartholomäus Steber

time, the culture thought nothing of sexual promiscuity among the upper classes, so there was no particular stigma to having a sexually transmitted disease. The specifics of how syphilis was treated were documented by a fellow named Ulrich Ritter von Hutten (1488–1523) who documented the horrors of his treatment process. Von Hutten wrote of enduring 11 mercury treatments over a period of nine years and then trying guaiac, which he said fully cured him. Lois Magner, the author of *A History of Medicine,* notes that his death within a few years may indicate that he wasn't as fully cured as he thought. Because syphilis was so unpredictable, there were examples to prove the "success" of every remedy.

A diagnostic blood test for diagnosing syphilis was created in 1906 by August von Wassermann (1866–1925). The bacterial cause of syphilis was not identified until the 20th century, and it took until then before any real progress was made against the disease. Salvarsan (an arsenical drug) was used before penicillin. Then in the 1940s penicillin began to be used effectively.

EARLY CONCEPT OF CONTAGION

The Italian physician and scholar Girolamo Fracastoro who named syphilis was a colleague of Copernicus at the University of Padua. Fracastoro taught medicine at several universities and also conducted very noteworthy studies. His work on contagion,

De contagione et contagiosis morbis (1546), was the first scientific writing that described the transmission of epidemics by transferable tiny particles or "spores" that could transmit infection. Fracastoro believed that each disease was caused by a different type of rapidly multiplying minute body, which were transferred from the infector to the infected in three ways: by direct contact, by carriers such as soiled clothing and linen, and through the air. Although microorganisms had been mentioned as a possible cause of disease by the Roman scholar Marcus Varro in the first century B.C.E., Fracastoro's was the first scientific statement of the true nature of contagion, infection, disease germs, and modes of disease transmission. His work attracted attention when it was introduced, but, because there was no science to move it forward, physicians more or less forgot about it until French chemist Louis Pasteur came up with germ theory in the 19th century.

FAMOUS RULERS THOUGHT TO HAVE HAD THE DISEASE

Several world leaders are now suspected of suffering from syphilis, and, because of its debilitating effects and its impact on brain function and personality, it may have affected history. Czar Ivan the Terrible of Russia (1530–84) became czar in 1547. Though he began his reign as a well-meaning leader, his children died at very young ages and his wife died in 1560. Ivan remarried and those children, too, were either unhealthy or stillborn. In 1564, Ivan's own behavior became erratic, and he exhibited symptoms that suggest he was suffering from cerebral syphilis. In 1565, he began ordering executions of people, and a 19-year reign of terror began. He and his sons raped the wives and daughters of those who were executed, and in 1581 he murdered his own son. He finally died in 1584. Many years later, his body was exhumed, and the speculative diagnosis was confirmed: Ivan had tertiary syphilis.

While it is often rumored that Henry VIII had syphilis, which would fit with his lecherlike image and murderous behavior, no one has ascertained that he actually had syphilis, and many feel that there are far too many other diseases that may have affected

Henry. Diabetes and circulatory problems—he suffered a series of strokes prior to death—are high among the other suggestions physicians give as to the illness from which Henry suffered.

PUBLIC POLICIES TO HELP REDUCE SYPHILIS

Public health matters generally had to be handled locally as there was no infrastructure on a higher level to set or enforce health policies. As syphilis began to invade various communities, however, town administrators knew that something needed to be done to reduce its spread.

With the understanding that it was a sexually transmitted disease, communities realized that the continued existence of brothels and prostitutes were at the core of the problem, and this was no small matter. Because a level of promiscuity was considered acceptable for the rich, Rome had 6,800 public prostitutes that could be accounted for at the end of the 15th century. The number would have been higher if there had been a way to count mistresses. Town officials began by attempting to expel nonresidents who were sick or to prevent them from entering at all, but then they began to create laws directed at prostitutes. Some cities expelled them to reduce the incidence of disease. However, in 1507 one town (Faenza) decided that the prostitutes did not need to leave, but they had to undergo health examinations in order to be certain they were not spreading the disease. (The nature of syphilis, with various stages and periods of dormancy was not well understood, so town officials inadvertently kept many prostitutes in business who were simply between stages of the disease.)

Anyone with active syphilis was in need of medical care, but physicians began barring prostitutes from hospitals out of fear of contagion. In 1498, in Bamberg, Germany, anyone with syphilis was forbidden to enter inns and churches or to have contact with healthy people. (The process of enforcing this was probably highly discriminatory.) In some towns, Wurzburg and Hamburg among them, special hospitals for syphilitics were created

U.S. STUDY OF SYPHILIS: A DARK CHAPTER

Syphilis was actually at the root of a bleak chapter in U.S. medical history. The U.S. Public Health Service started a medical evaluation in 1932 that was based on a study used in Oslo that patched together information about the course of untreated syphilis. The U.S. study is now viewed with horror and embarrassment.

At the time, standard medical treatments for syphilis were widely known to be toxic, dangerous, and not necessarily effective. The idea behind observing syphilis untreated was twofold: to determine if patients did better without the toxic "cures," and to identify the stages of syphilis with the idea that a stage-specific treatment might be effective. To buttress what the Swedish researchers were doing with a retrospective study, scientists in the United States chose a county—Macon County in Alabama—with a very high rate of syphilis and also a high rate of poorly educated African Americans. Six hundred poor black men were put into the study conducted by the Tuskegee Institute and the Veterans Administration with promises of free medicines, regular medical care, burial assistance, free hot meals on the days of examination, transport to and from the hospital, and sometimes an opportunity to shop while in town. Of the group, 399 were thought to have syphilis and 201 were in the control group.

The study group was frequently misled as to what was happening to them; "treatments" were often nothing more than placebos, and spinal taps for evaluative diagnosis were billed as "Last Chance for Special Free Treatment." Then in 1947 another boundary was crossed when penicillin began being used effectively against syphilis. The decision was made to continue the study and also to continue to withhold

(continues)

(continued)

penicillin from the men without telling them that an effective treatment had been found.

The study was not a secret. Results were published regularly, but it was not until 1966 that the study was called into question. Peter Buxtun, a Public Health Service investigator, filed an official protest with the Division of Venereal Diseases of the Centers for Disease Control. His objection was ignored since the study was not yet complete. (The study was to be considered complete when all participants had died.) He raised the issue again in 1968 and was ignored, so in 1972 he leaked the story to the *Washington Star.* The reporter Jean Heller's story appeared on July 25, 1972. Other newspapers picked it up, and the study was quickly brought to an end in November of 1972 when the press turned public sentiment against the methodology. By this time, 28 men had died of syphilis, another 100 had died of complications related to syphilis, at least 40 wives had been infected, and 19 children had contracted the disease at birth.

The lawsuit against the study was settled out of court with each survivor receiving $37,500 in damages and the heirs of the deceased receiving $15,000. Since that time, the government has created a method to evaluate its research practices and to monitor all studies using human subjects.

by municipalities. Free care was often provided, but as part of the deal physicians were legally obligated to report who had the disease.

Over time, people seemed to develop some level of immunity or the disease became somewhat less ravaging; syphilis became more of a chronic problem, and people did not die from it as quickly.

As the Renaissance waned and middle-class morality began to exert more influence, more social stigma became associated with venereal diseases, including syphilis. This sent the disease underground, which complicated efforts at treating it.

CONCLUSION

Current knowledge indicates that syphilis did not appear in Europe until the end of the 15th century, and its occurrence, the attitude toward it, and its treatment was indicative of medical care of the day. Because there was a more open attitude about extramarital sexual activity among the wealthy, the root cause of the illness could be dealt with more directly, though not necessarily particularly effectively. The fact that syphilis went through various stages with periods of dormancy made it difficult for physicians to judge what treatments were helpful. The determination that mercury was the cure-all was a destructive philosophy that made people sicker, and the error of this thinking was not to be discovered for almost 450 years.

7

The Impact of the New World on Medicine

The field of botany and pharmaceutical medicines was one of the areas most affected by the fact that voyagers were beginning to travel all over the world. As explorers set off for parts unknown, they returned with dried plants and seeds to grow new plants that led to new discoveries. Some of the discoveries were very helpful—as Peruvian bark was with malaria—and some were thought to be helpful, like tobacco, but later proved to be harmful.

Prior to the arrival of the Europeans, there were few diseases that affected the Native Americans. The health issues they needed to deal with were generally injury, digestive disorders, and rheumatism for those who lived into older age. Native Americans were very knowledgeable about the use of plants for nutrition and medical purposes. Scientists today verify that many of their treatments would have been effective. For example, the ingestion of a powder from the white willow tree (*Salix alba*) would have been much like aspirin. The ashes from the tree could also be used for sore eyes, and the interior wood was mashed and used as an antibiotic for wounds. Lobelia (*Lobelia inflate*) could be smoked and caused bronchial tubes to dilate, which helped with asthma, bronchitis, and whooping cough. It was also a narcotic. This type of informa-

A few items unearthed at Jamestown that were used by doctors and apothecaries, including drug jars, ointment pot, bleeding bowl, mortar and pestle fragments, glass vials, and portions of surgical instruments *(Project Gutenberg)*

tion about plants became the building blocks of modern pharmacology. (Most of the information about Native American practices was carried down by oral tradition; only the Aztecs recorded things, and their records were largely destroyed by the Spanish.)

Medicine has always been more available to the wealthy than to the poor, and one English physician Nicholas Culpeper set out to right this wrong. He had trained under an apothecary and was a serious student of astrology so his field of specialty combined the use of medicinal plants and keeping a close eye on how the stars might affect what he prescribed.

This chapter explores Native American lore about medicine and the types of plants they used. The arrival of Europeans in places where the Native people had no immunities was disastrous for the Natives, and what happened to them will be discussed. Nicholas Culpeper was a healer who became intent on helping the less fortunate, and why he did what he did will be explained.

THE NEW WORLD INFLUENCES MEDICINE

To understand how New World medicine came to have an impact on European culture, it is important to consider who the first arrivals were and what challenges they faced. When the first group of Englishmen—and it was all men and boys—were funded by King James I to go to the New World and create a settlement, they were sent with two primary goals: They were to find gold for the king and locate a water route to the Orient.

The men were woefully unprepared for both the challenges of the journey and those of the destination. The voyage itself took 26 weeks, and by the time they arrived in what they would later call Jamestown (after the king), many of them were quite ill from diseases that resulted from poor nutrition during the crossing (primarily beriberi and scurvy).

The group consisted of 39 crewmen and 104 other travelers, the majority of whom were gentlemen. Some knew a little about agriculture, but few were knowledgeable about construction, and there were no physicians onboard.

The London Company, corporate parent of the Virginia Company, assumed that any health issues that cropped up would be injury related, so on subsequent voyages some barber-surgeons, apothecaries, and healers were sent, but no one really thought of sending a university-trained physician. The London Company understood the importance of a healthy population; they knew that without health there was no way to make money from the settlement. The only issue was that they did not foresee all that was to befall the colonists. (Anyway, well-regarded physicians would have preferred traveling to Europe. They valued the European universities, and the opportunity to teach at another university or to fraternize with other professionals would have been much preferred than to be sent with a band of colonists to an unknown land.)

The challenges were enormous, and the men simply were not up to the task of creating a new life for themselves in the wilderness. They were so intimidated by the Native Americans that they were reluctant to farm the land outside the fort, and they had soon

Typical Apothecary of the Renaissance

Elaborate concoctions were still used in medicine, so apothecaries featured bottles that were easy to store and from which to pour.

hunted most of the available animals in the immediate area. Some turned to cannibalism, and others simply died of starvation. By the end of the first year, only 38 of the original group survived.

Captain John Smith was one of the first to notice that the Native Americans were in notably fine health, and when additional ships arrived he was intent on taking items from the ship that could be traded with the Native people for maize (corn) to feed his people or for advice on what to do about illness.

In 1610, Lord Delaware, the colonial governor at the time, brought with him Dr. Lawrence Bohun, a very well-regarded physician and surgeon from the Netherlands. Dr. Bohun noted the new plants available and began investigating them in order to supplement the medicines the colony had on hand. Bohun, however, still believed in the humoral balance. As it turned out, the community was essentially doomed. Between 1607 and 1624, 6,000 people came to Jamestown. By 1625, only 1,200 remained, and they were not in good health.

WHAT THE NATIVE AMERICANS KNEW

Native Americans spanned the continent, and there were many different tribes even within general geographical areas. Acknowledging that each area of the continent and each tribe had unique approaches to maintaining good health, a few generalizations can be made. Most tribes were nomadic so they moved about as they needed. This meant that sanitation was rarely an issue; they moved on often enough that they would rarely have had to face what to do about water they might have polluted with their waste. They all pursued herbal cures from the plants that grew in their areas, and most turned to medicine men or medicine women (also called *shamans*) to guide them on proper treatment. Spiritual belief was a major component of their existence, and they believed that sickness occurred when the spirits were displeased. The job of the shaman or medicine man was to use herbs and prayers—often in the form of dances, chants, and incantations—to win over the deity that was offended. In most tribes, the job of medicine man or woman was prestigious but it also carried a great deal of pressure. If the tribe's luck turned or the medicine man had a bad string of failed cures, he would be killed or driven from the tribe.

Some tribes had more than one layer of healer. Some had herbalists who would have been the first to be consulted; next was a hand trembler or diviner who would have stepped in if the herbal cures did not work. Finally, the medicine man would be consulted. Medicine men carried medicine bags filled with various plants, amulets, and ingredients for concoctions that could be mixed depending on what was needed. (Some tribes believed every young man should carry a medicine bag, and theirs was an elaborate ritual preparing the bag that each brave carried for his lifetime.) The bags were always made of cured animal skins that were decorated with symbols to bring good health and good luck.

Native Americans often took a very scientific approach to their medicines. They had worked with these plants and herbs for a long time and knew there were preferable times for planting, for digging certain roots, and for maximizing benefits from certain flowers or leaves. They generally turned the herbal cures into two types of treatment—*decoction* that involved boiling a plant in water, or *infusion* that involved boiling the water and removing the solid that remained after it cooled.

TRADE AFFECTS BOTH SIDES

Native Americans received two disastrous "imports" from the Europeans. Infectious diseases, as many historians have noted, did much of the "conquering," because the Europeans arrived with illnesses to which the Native Americans could not possibly have had any immunity. (See section about smallpox in chapter 5.) Smallpox, measles, tuberculosis, *scarlet fever,* diphtheria, typhoid, and malaria may all have been introduced by the Europeans.

The second disastrous import that the Europeans brought with them was "medicine water"—whiskey. Native Americans soon learned that whiskey "made the pain go away," but this was to have a devastating effect on them throughout the continent for generations to come.

MEDICINES FROM OVERSEAS

As explorers, missionaries, and colonizers returned from their voyages, they brought with them a variety of plants that began to have a major impact on therapeutic treatments of the 16th and 17th centuries:

- Peruvian bark (*Cinchona officinalis*) came from South America between 1630 and 1640 and was said to help bring down fevers. Jesuit priests began recommending it, so it was also known as Jesuit's bark. (A legend of how it cured Countess Anna del Cinchon, wife of the Spanish viceroy of Peru, was part of the lore surrounding the medicine, and her name is at the root of the Latin name for the plant.) Peruvian bark was added to the *London Pharmacopoeia* in 1677 and because it was highly effective against some illnesses, it became widely demanded for fevers of all types. It was difficult to import as much as was in demand, however, and the search was on for sources of the bark. Later quinine was noted to be contained in the plant, and it continues to be a drug used for treating malaria.

Peruvian bark (*Cinchona officinalis*) was brought back from the Americas and proved beneficial for treating diseases such as malaria.

- Opium was primarily brought in from the east. (See *Opium as a Medicine*.)
- The ipecacuanha plant (*Cephaelis ipecacuanha*) plant was brought back from Brazil where the Europeans learned that it could be a powerful medicine. The dried root of the plant could be used to stop certain types of diarrhea (particularly amoebic dysentery); ipecac also served as an effective *emetic* in some cases of poisoning. If taken in even smaller doses, it could be used as a cough *expectorant*. Modern first-aid kits include ipecac, though families are to consult poison control centers before using it, because some poisons should not be regurgitated.

The Ipecac Plant

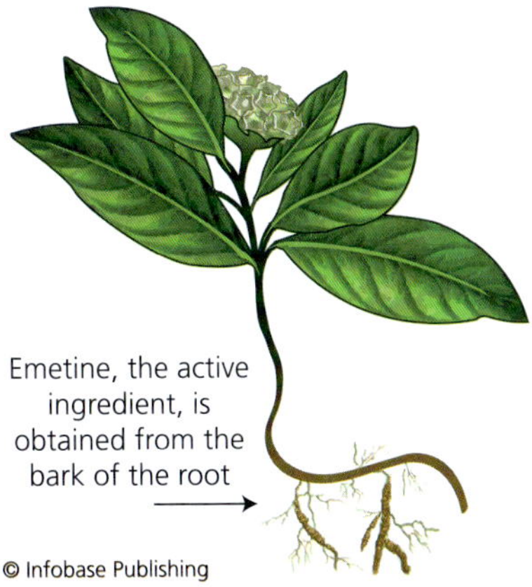

The ipecac plant was imported from the New World and used in various treatments.

- Tobacco was another plant that was thought to be medicinal. Tobacco is native to North and South America. It is part of the same family as the potato, the pepper, and the poisonous nightshade. The Aztecs smoked hollow reeds stuffed with tobacco leaves, and Central and North American natives smoked leaves that they wrapped in palm leaves or maize husks. When Christopher Columbus arrived in what he thought was the Indies, the natives communicated that the herb was used for medicinal purposes. In the 1550s, sailors took tobacco back to France and Spain. Sir Walter Raleigh is thought to have been the first to bring it to England in 1565. During the 16th and 17th centuries, sailors used it throughout South America, the Caribbean, and North America. Tobacco was seen as a cure-all when it was first imported. It was recommended for toothache, poisoned wounds, joint pains, bad breath, ulcers, *chilblains,* tiredness, and much more. During the Great Plague of 1665, it was commonly believed that smoking could guard against illness, and schoolboys at Eton in England were flogged for not smoking often enough. There were some early naysayers. One of the most prominent was King James I of England. He described it as a "stinking loathsome thing," and he tried to limit tobacco supplies by raising excise duties. In the process, he encouraged a huge black market in

(continues on page 116)

OPIUM AS A MEDICINE

Opium is a highly addictive drug that is derived from the poppy plant, *Papaver somniferum.* (*Somniferum* is a Latin word that means "I bring sleep.") Raw opium is dark brown and gummy with a very strong odor and a bitter taste. Those who consume a small bit of it (50 mg) gain a sense of well-being; those who take larger doses can die.

Today, most poppy-harvesting takes place in the Golden Triangle (Laos, Burma, Thailand) the Golden Crescent (Afghanistan, Pakistan, and Iran), and Mexico. The opium is collected from the poppy capsule that is essentially the fruit of the flower after the poppy blooms and the petals fall off. (A single poppy plant can have five to eight poppy capsules.) To collect the opium, the capsule must be lanced, which involves making a shallow incision in the capsule. Each capsule is hollow but contains several chambers called loculi that contain thousands of tiny, kidney-shaped seeds. The incision is deep enough to lacerate the *laticiferous* vessels of the capsule so that the latex can begin to ooze out, a process that takes several hours. The timing of the incisions must be precise so that wind or rain does not affect the exudation. Generally, the next morning the latex is scraped off with a knife. When opium is being

Opium Plant

The highly addictive drug opium is harvested from the poppy flower.

harvested, this process is repeated several times over two to three days.

Raw opium consists of several special chemicals known as *alkaloids,* which are bitter-tasting chemicals. All alkaloids are poisonous, but if taken in very small doses they can work as drugs. (Alkaloids are recognizable by name because in English they end in "ine," such as nicotine or strychnine, or cocaine.) The bitter taste may have been designed by nature to warn off animals. The primary alkaloid in opium is morphine, a potent suppressor of pain. (The word *morphine* comes from the Greek Morpheus, "god of dreams.") Morphine was not isolated from opium until 1805, but opium itself has been used since ancient times.

Opium has been around for an exceedingly long time. While opium is referred to in clay tablets as early as 5000 B.C.E., opium was being used frequently in Egypt to the point that reports were that people became faint from want of it. Opium is mentioned in the *Ebers Papyrus* where it was described as a good thing to use to quiet children, and mothers used to rub poppy juice on their nipples to help nursing babies go to sleep. Greek soldiers were said to take nepenthes, a drug made from opium, and they tended to take it before going into battle to dull their sense of danger. And no less a personage than Galen touted the virtues of opium.

Besides being a very strong suppressor of pain, opium suppresses cough and produces constipation, thus being very useful in cough and diarrhea. In fact, a form of opium known as *laudanum* (from the Latin word *laudare,* meaning "to praise") became very popular in the 17th century for treating *dysentery.* The British physician Thomas Sydenham (1624–89) (see chapter 8) virtually put an official stamp of

(continues)

(continued)

approval on opium by advocating its use in dysentery and other such conditions. Also known as *tincture* of opium, laudanum was nothing but a solution of opium in alcohol (10 percent opium or 1 gm of morphine to 100 cc of alcohol). Sydenham flavored the tincture with saffron, cinnamon, and clover. This exotic preparation came to be called Sydenham's laudanum and became a very popular remedy in Europe. So enthusiastic was his advocacy of opium that Sydenham won the nickname "Opiophilos" (lover of opium). In 1680, Sydenham wrote: "Among the remedies which it has pleased almighty God to give to man to relieve his sufferings, none is so universal and so *efficacious* as opium." His pupil, Dr Thomas Dover (1660–1742), invented the famous Dover's powder, which contains 10 percent opium. Dover's powder became a popular remedy for alleviation of pain and cough.

Almost a century before Thomas Sydenham introduced opium in laudanum, the Swiss physician Paracelsus referred to opium as the "stone of immortality." He was an opium-eater himself. He once boasted, "I possess a secret remedy which I call laudanum and which is superior to all other heroic remedies."

(continued from page 113)

tobacco and, realizing this, he changed his mind. He also came to understand that the new drug could be imported from the new British colony of Virginia, and, in so doing, he could make a lot of money.

As time went on, London and Amsterdam became important markets for pharmaceuticals and, by 1750, *pharmacists* in both cities were shipping different medicines worldwide.

HEALTH CARE FOR THE COMMON MAN

Nicholas Culpeper (1616–54) was a practicing English physician who had become fascinated by botany during his childhood. Some unexpected twists of fate led Culpeper into his profession as the champion of medicine for the common man, and today he is known for the work he did in herbal medicine. Some books refer to him as the father of alternative medicine. However, it is important to remember that medicine was not yet a real science, so although Culpeper may have been practicing differently from other physicians, it is hard to define what was alternative.

Culpeper was born into a well-to-do family, and his father, a minister in Surrey, England, died before Culpeper was born. He was brought up in the home of his stern and strict Puritan grandfather. As a boy, Culpeper read widely from his grandfather's well-stocked library and was particularly interested in astrology and plants. Among the books he read was William Turner's *Herbal* (1568), which was to make an impression on him. Culpeper's grandfather wanted the young man to attend Cambridge University as he had and then to become part of the ministry as his father had done.

In accordance with his grandfather's plans, Culpeper enrolled in Cambridge, but he did not enjoy theology and added classes on medicine and astrology to his course load. With plants being discovered in the New World, smoking was becoming fashionable, and Culpeper took up the habit.

Culpeper's childhood sweetheart and his mother died within a year of each other (his sweetheart was struck by lightning; his mother died of breast cancer). Culpeper was devastated and refused to return to Cambridge. Greatly disappointed, his grandfather disinherited him but saw that he was placed as an apprentice to a master apothecary. Apprenticeships at that time generally took seven years. Culpeper was happy in the apprenticeship and also encountered one of the most important astrologers of the day, William Lilly (1602–81), so Culpeper was able to indulge in both fields of study. More bad luck hit, however, when the apothecary for whom he was working went bankrupt and left the country.

Culpeper attached to a second apothecary, but this fellow died before Culpeper had completed the seven-year apprenticeship. At this point, he and a fellow apprentice took over the practice under the supervision of Stephen Higgins, master of the Society of Apothecaries. Higgins took them on plant-gathering excursions and taught them what was known about medicinal herbs and plants.

In 1639, Culpeper married a wealthy young woman whose father he had treated for gout, and with the financial security he now enjoyed he decided to set himself up as an astrologer, herbalist, and physician. In his practice, he combined herbal remedies with what he had learned about astrological influence as well as some of the medical understanding of the day. He located his shop in the poorer section of London's East End as he wanted to help those who did not have easy access to medical care. His move was met with great anger by the Society of Apothecaries. Because he had not fulfilled his full seven-year term, they saw this as flouting their authority. Despite this, Culpeper gained a favorable reputation among the poor in the area. He never turned anyone away and often saw as many as 40 patients a day, charging little for his services.

In 1642, Culpeper responded to the need for soldiers for the English Civil Wars. However, when they learned he was a healer, the officers took him along with units that needed field surgeons. He eventually was placed as captain of an infantry. In 1643, he was hurt, with an injury that was to bother him for the rest of his life.

When he returned to London, the richer for the medical knowledge he had acquired, he determined that he needed to speak up for the common man. Culpeper believed that the College of Physicians and the Society of Apothecaries so dominated the profession that it kept the cost of medical care high and used ingredients in the medicines that were pricier than necessary. (A charter of the Society of Apothecaries in 1617 set out very specific guidelines for mixing medicines. It was written in Latin and employed only the more costly drugs and medicines.)

During one of his apprenticeships, Culpeper had begun to translate this pharmacopia from Latin into English, but it could not be published in any other form because the Company of Stationers had been charged with censoring anything that was against the establishment (1603). Offenders were fined, imprisoned, whipped, or mutilated. In the 1640s, during the early part of the Civil Wars, some of these regulations were abolished, and, in 1649, Culpeper published *A Physical Directory, or a Translation of the London Directory*. Culpeper's action enraged the College of Physicians who attacked him in the press, but Culpeper maintained his right to publish. He went on to write *Directory for Midwives* (1651). Death during childbirth and infant mortality were common at this time, and Culpeper and his wife had seven children, only one of whom survived infancy.

Culpeper continued prescribing medicines mixed with less expensive herbs, and he published several more books that combined his astrological beliefs with his knowledge of herbal remedies. The astrological ideas from that time were quite similar to the belief in balancing the humors. Astrology was discussed in terms of the four elements—earth, water, air, and fire—and everything was classified by whether it was hot, cold, moist, or dry.

Culpeper's two most famous works were *The English Physician* and *The Complete Herbal*. The books were written in English (and all plant names were in English rather than Latin), and Culpeper wanted them priced inexpensively so that they could be available to all. Culpeper believed that medicine was a public asset and should not be treated as a commercial secret.

Culpeper died in 1654 at the young age of 38. His health had not been good since suffering his war injury. Some felt he contracted tuberculosis because he was weakened by this. His secretary, W. Ryves, noted that his smoking was a contributing factor: ". . . the destructive Tobacco Mr. Culpeper too excessively took, which by degrees first deprived him of his stomach and after other evil effects in the process of time, was one of the chief hasteners of his death."

Culpeper's two most prominent works were in active use for 250 years after his death, and, while some historians are critical

Some Medicinal Plants of the Day

Plant	Historical uses
Bergamot	Colic, fevers, nosebleeds, gas, heart trouble, to expel worms
Bloodroot	Fever, rheumatism
Boneset/Comfrey	Broken bones, sprains, bruises, diarrhea, dysentery
Butterfly Weed	Respiratory ailments, stimulant
Calamint	Soothes indigestion
Catnip	Headache, stomachache, insomnia, bronchitis
Dusty Miller	Cataracts
Elderberry	Diarrhea, purgative, colic, ointment, lotion
Elecampane	Expectorant for coughs
Foxglove	Heart stimulant
Garlic	Colds, fever, cough, bronchitis, high blood pressure, headache, rheumatism
Ginseng	Tonic, stimulant
Hops	Sedative, antispasmodic, diuretic, insomnia, cramps, cough, fever
Lamb's Ears	Wounds
Lavender	Headache
Lungwort	Coughs, pulmonary problems
Monkshood	Toothache, neuralgia, sciatica
Parsley	Laxative, kidney problems, jaundice
Peppermint	Indigestion, colic, intestinal ailments, ointment
Rosemary	Astringent, induce sweating, headache, hair rinses
Sage	Stimulant, antiseptic on wounds
Snowdrop	Pain reliever, dysentery, stimulate menstrual flow
Sweet Woodruff	Wounds
Tansy	Insect repellent, colds, stomachache, sprains
Violet	Pulmonary problems, pain reliever, inflammation
Wild Rye	Opthamolic
Yarrow	Indigestion, gastric inflammation, internal bleeding

Note: Historical information only; many of the plants listed are toxic

This chart shows some of the medicinal plants of the day and how they were used.

of his work, he took an important stand for the common man. In a day when a regular physician was likely to suggest bloodletting, patients were in adequate and caring hands when they consulted astrologer and herbalist Nicholas Culpeper.

CONCLUSION

The field of plant-based medicines was one of the areas most affected by voyagers traveling the world. As explorers set off for parts unknown, they returned with dried plants and seeds to grow new plants that led to new discoveries. Some of the discoveries were very helpful—as i.e., Peruvian bark with malaria—and some were thought to be helpful, such as tobacco, but were later proved to be harmful. Others such as opium may have been helpful in the short term, but the long-term dangers of addiction were not understood.

Nicholas Culpeper made major contributions in the field of medicine by realizing and standing up for the fact that medicine should be available to both the rich and the poor, and he did all he could to stand up for the rights of the underprivileged.

8

Scientific Progress on an Imperfect Path

As so many aspects of life progressed, medical science was also able to move forward. One helpful trend was that phenomena could be dealt with mathematically. Scientists and physicians began to see that it was productive to keep quantitative track of life trends. This dovetailed nicely with a renewed interest in studying disease. Several diseases were described for the first time—among them were whooping cough, typhus fever, and scarlet fever. However, just because scientists were learning new truths, it did not necessarily mean that medical care was getting better.

As populations grew and lived in more densely settled communities, scientists began to understand that there needed to be new ways to deal with waste and to provide clean water, but that these were not easy problems to solve.

This chapter introduces Thomas Sydenham, known as the English Hippocrates because of his belief in observation-based medicine. Diseases were just beginning to be studied for their possible relationship to certain occupations, and this chapter concludes with a focus on the state of sanitation and public health during Sydenham's time.

THE ENGLISH HIPPOCRATES

During his lifetime, Thomas Sydenham (1624–89) enjoyed a reputation as a successful physician who helped his patients feel better. Over time, however, as scholars studied his methods, his reputation grew to the point that he is sometimes referred to as the English Hippocrates or the Father of English Medicine. His approach to patients revived the Hippocratic technique of careful observation of patients and basing his treatment on what he observed. He also kept careful and detailed records about each patient, so he has become known as a founder of clinical medicine. He also is credited because of the groundwork he laid for epidemiology because he undertook careful studies of various *epidemic* illnesses ranging from smallpox to scarlet fever. Sydenham introduced laudanum, was one of the early practitioners of iron use in treating anemia, and popularized the use of cinchona (quinine) in treating malaria.

Sydenham was born into a well-off family in Dorset, England. His education at Magdalene Hall, Oxford, studying medicine was interrupted by the need to join the military during the English Civil Wars. He returned to Oxford in 1645 to continue his education in medicine. At Oxford, he met many of the scientists who formed the Royal Society, but after approximately 18 months of education he again rejoined the army (1651) and continued service until 1663 when he married and opened a practice in London.

Most practitioners of Sydenham's time believed heavily in a theoretical approach to medicine, followed by experiments of various types. Sydenham's reintroduction of Hippocrates' patientcentric beliefs was novel for the time. Scholars feel that his time in the military may have formed his philosophy of how important it was to treat patients based on bedside observation rather than theory. An oft-quoted saying of Sydenham's noted that the art of medicine was "to be properly learned only from its practice and its exercise."

The chemist Robert Boyle, whom Sydenham got to know at Oxford, encouraged him to study the nature of epidemics, and Sydenham's first writings was a book on fevers that was

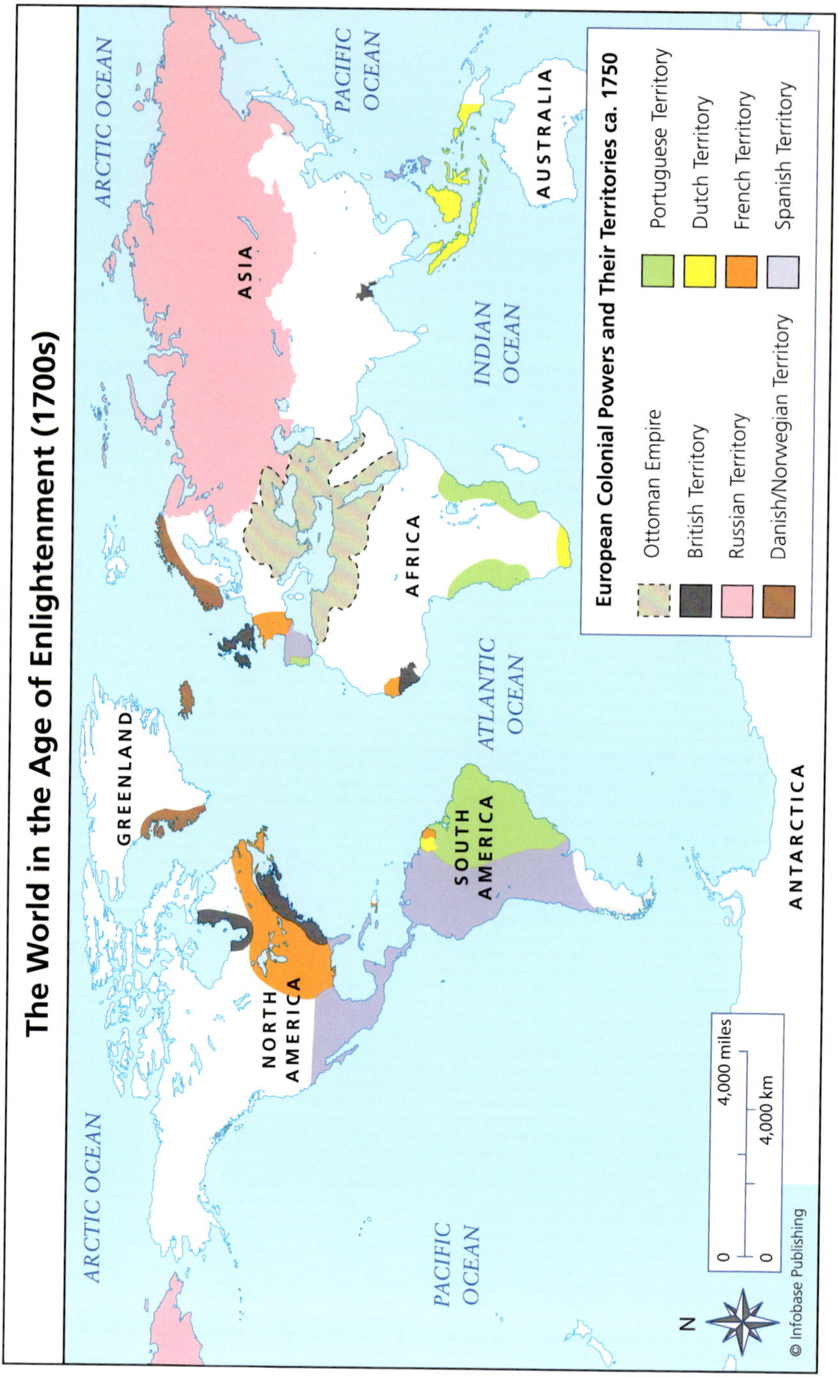
The World in the Age of Enlightenment (1700s)
ARCTIC OCEAN
PACIFIC OCEAN
ASIA
INDIAN OCEAN
AUSTRALIA
AFRICA
ATLANTIC OCEAN
GREENLAND
NORTH AMERICA
SOUTH AMERICA
ANTARCTICA
ARCTIC OCEAN
PACIFIC OCEAN
European Colonial Powers and Their Territories ca. 1750
Ottoman Empire
British Territory
Russian Territory
Danish/Norwegian Territory
Portuguese Territory
Dutch Territory
French Territory
Spanish Territory
0
4,000 miles
0
4,000 km
N
© Infobase Publishing

published in 1666. Between 1669 and 1674, Sydenham kept a detailed notebook of clinical observations, and this was the basis of his major work, *Observationes medicae circamorborum acutorum historiam et curationem* (*Observations of medicine*), which was published in 1676.

Sydenham suffered from kidney stones and gout, so these two illnesses benefited from his detailed descriptions, but he also described other disorders for the first time, including measles and scarlet fever. Saint Vitus' dance is also known as Sydenham's chorea, still bearing his name. He drew a connection between mosquitoes and typhus, and this was revolutionary for the time. He was respected for the cooling regimen that aided smallpox or any type of fever, and his use of laudanum and cinchona were helpful in treating various illnesses.

Because Sydenham was out of step with his contemporaries, he died as a well-respected but not highly praised practitioner. His reputation grew over time as physicians began to follow the precedents he set.

ALCHEMY WANES: IDEAS SUCH AS PHRENOLOGY TAKE ROOT

Men who followed Sydenham's ideas were beginning to help society move away from the belief in magical cures, and, while some alchemists provided the thinking that led to the early foundation of chemistry, some of them were still rooted in what was essentially a process of trying to transform metal into gold, or in the world of medicine, "nothing into something helpful." But change came slowly, and some aspects of alchemy were to live on for another 100 years.

As the science of anatomy became more important, some physicians became interested in figuring out how to fix parts of the

(Opposite) By the 1700s, Europeans were traveling the world and bringing back new plants that were being mixed and sold as medicines.

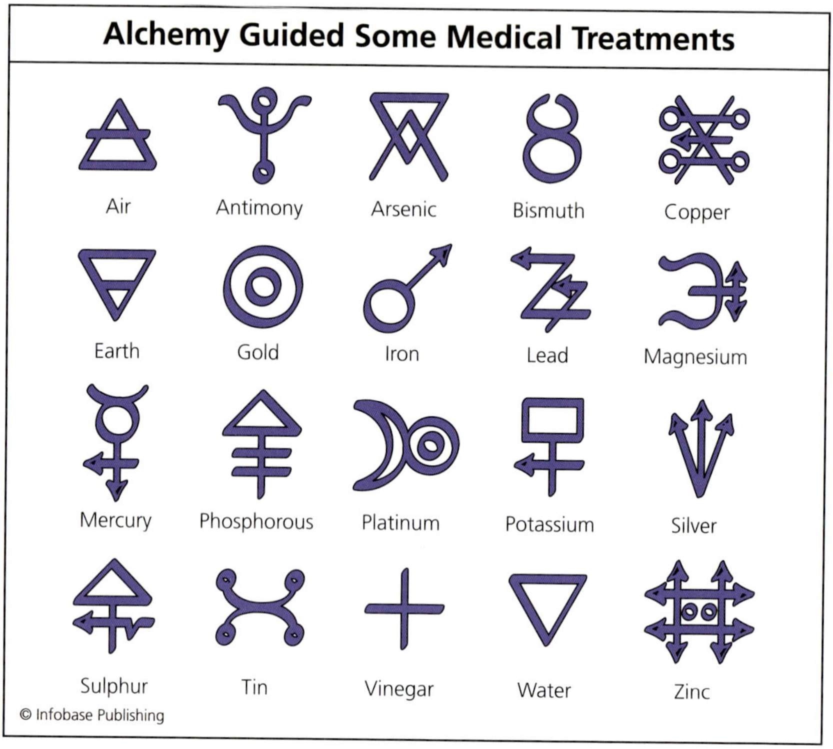

These were the symbols that an alchemist used in his work.

body. While the science of phrenology was not to be fully developed for another 50 years, practitioners were beginning to try to understand the workings of the body and how the brain did or did not control certain things.

CONNECTING CERTAIN JOBS TO CERTAIN DISEASES

While scurvy (see chapter 5) occurred in many poor areas and among travelers who took part in the Crusades, it was a particular problem among seamen who were away on long voyages. Other groups of workers suffered from the same health problems, and physicians and scientists were beginning to make the link between employment and illness. In 1556, Georgius Agricola (1494–1555), a German scholar and scientist, wrote a treatise on diseases common to miners. As early industry created a greater call for minerals, it necessitated digging deeper mines, and that created problems from breathing in mineral-laden dust. Problems were not just seen in miners. Eleven years after Agricola's book, Paracelsus (see chapter 1) wrote a three-volume work that marked the beginning of occupational medicine (1567). Each volume addressed a different topic: The first talked of diseases common to miners (mainly pulmonary problems); the second addressed diseases of smelter workers and metallurgists; the third talked of diseases caused by mercury. In addition to tracing the causes of the diseases, Paracelsus also wrote of prevention, diagnosis, and therapy. While no helpful discoveries were made that helped miners during this time, the growing awareness was an important issue that permitted others to pursue these studies more seriously.

Goldsmiths were also among those noted as suffering from similar illnesses. A physician who wrote about this was Ulrich

A Phrenologist's View of the Brain

1. Physical love
2. Parental love
3. Attraction to others
4. Combativeness
5. Destructive tendency
6. Secretiveness
7. Acquisitiveness
8. Hunger/appetite
9. Mechanical ability
10. Awareness of time
11. Awareness of size, color, order
12. Individuality
13. Awareness of facts/circumstances
14. Analytical ability
15. Understanding of cause and effect

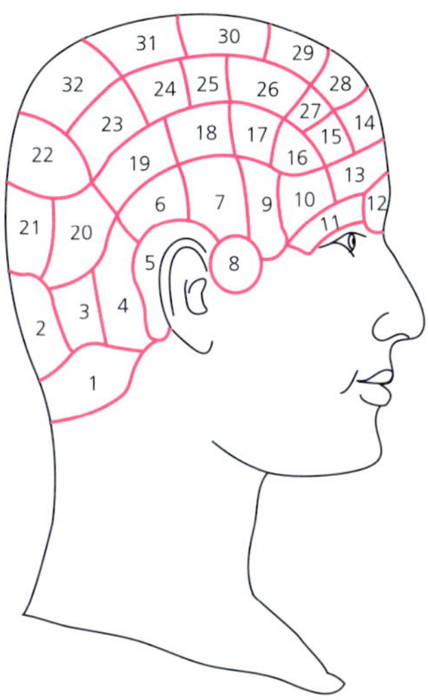

16. Humor
17. Good taste
18. Appreciation of the magnificent
19. Caution
20. Friendship
21. Love of home
22. Ability to concentrate
23. Ambition
24. Conscientiousness
25. Hope
26. Faith
27. Pleasantness
28. Sympathy
29. Agreeability
30. Devotion/respect
31. Firmness
32. Dignity

Scientists were just beginning to explore the brain.

The Quack by Franz Anton Maulbertsch *(The Yorck Project)*

Ellenbog, who wrote *Von den gifftigen besen tempffen und reuchen (On the poisonous, evil vapors and fumes of metals, such as silver, quicksilver, lead and others which the worthy trade of the goldsmith and other workers of metals are compelled to use: How they must conduct themselves to dispel the poison.* 1523 and 1524). He also offered preventative advice.

This led the way for Bernardino Ramazzini of Modina (1633–1714) who wrote the classic *De morbis artificum diatribe (Discourse on the diseases of workers)*, a work published in 1700. He eventually contributed works that outlined the health hazards of chemicals, dust, metals, and other agents encountered by workers in 52 different occupations. The first edition of the book addressed issues involving miners, gilders, apothecaries, midwives, bakers, millers, painters, potters, singers, and soldiers. A 1713 edition added 12 more groups: among them printers, weavers, grinders, and well-diggers. Ramazzini approached the issue on two levels. He reviewed what was known or observed at the time among various professions, and he also set the stage for what needed to be considered. His book was translated into many languages and used as an important text on occupational illnesses until the 19th century when everything changed because of the Industrial Revolution.

THE FOUNDATIONS OF PUBLIC HEALTH

During the 16th to 18th centuries, public health continued to be handled locally. Those who headed national governments were fixated on expansion and power, looking for whatever might give them wider control over the world. There was acknowledgment that a large and healthy population contributed to a wealthier, stronger country, but the amount of effort that larger government entities put into controlling disease was minimal. National governments did not have the resources or the advisers with knowledge to create any sort of workable health platform.

One advance in public health that was to be wide-ranging was the advent of studying public health mathematically.

Early Brass Vaccination Syringe

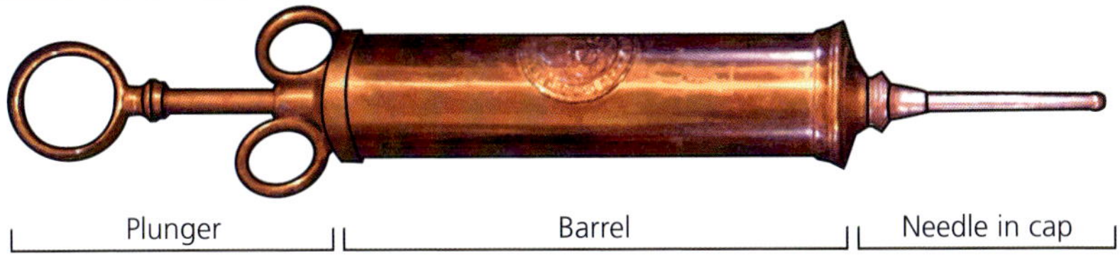

Physicians were just beginning to experiment with vaccinations. Most were not effective.

DOCTORED TO DEATH

The physician Sir Charles Scarburgh (1615–94) is significant today because he left behind a manuscript describing the manner in which King Charles II was treated for his final illness in 1685, and that document has been preserved at the Society of Antiquaries in London and has been accessed by scholars. Today, it is sometimes said he was "cured to death."

Scarburgh was a physician who taught at Oxford for many years before becoming a physician to the king. The king was being shaved the morning of February 2, 1685, when he suffered a convulsion. Although Scarburgh was called in, he also gathered 12 more physicians to advise him because he did not want to be solely responsible for treatment of such an important personage.

The first thing they undertook was to bleed Charles, taking a pint of blood from his arm. Then an incision was made in the king's shoulder and another eight ounces of blood was removed through a cupping process. Emetics and purgatives were given, then a second purgative, and then an enema that contained "antimony, sacred bitters, rock salt, mallow leaves, violets, beet roots, chamomile flowers, fennel seed, linseed, cinnamon, cardamom seed, saffron, and aloes." This process was repeated two hours later and then another purgative was given. Next, Charles's head was

Governments began to do political arithmetic that provided data that had never before been gathered, and this was to lay an important foundation in mastering a better understanding of public health. William Petty (1623–87) is considered the father of political arithmetic. He was an English physician, economist, and scientist who believed that a healthy country needed a healthy shaved, and a blister was raised. He was then given a sneezing powder of hellebore root and another powder of cowslip flowers "to strengthen the brain." Cathartics were given frequently, and then a soothing drink (barley water, licorice, and sweet almonds) was given. White wine, absinth, and anise with extracts of thistle leaves, mint, rue, and angelica were also administered. His feet were then covered with a plaster of burgundy pitch and pigeon dung. More bleeding and purging was done, and medications containing melon seeds, manna, slippery elm, black-cherry water, extract of flowers of lime, lily of the valley, peony, lavender, and dissolved pearls were given. After this, gentian root, nutmeg, quinine, and cloves were given. Scarburgh reports that the king's condition worsened so 40 drops of human skull were prescribed to prevent more convulsing. Then an antidote containing herbs, animal extracts, and Bezoar stone was administered.

Scarburgh then noted: "After an ill-fated night his Serene Magesty's strength seemed exhausted to such a degree that the whole assembly of physicians lost hope and became despondent; so as not to appear to fail in doing their duty in any detail, they brought into play the most active cordial."

It was noted that the king was unconscious during most of these ministrations, and of course the final result was death.

population, and he began to collect data on population, education, diseases, revenue, and other topics. Petty spoke out about how important it was for the state to foster medical progress, and he also advocated that hospitals should help train physicians. In addition, he proposed a health council for London to deal with public health matters (1687).

John Graunt (1620–74) was a haberdasher who became an early demographer at the encouragement of his good friend William Petty. Graunt started collecting data and went on to write a classic book *Natural and Political Observations . . . upon the Bills of Mortality,* which was first published in 1662. Graunt looked back over a 25-year period in London and began noting the number of deaths in London during the preceding one-quarter of a century. He noted the gender, the age, the place of residence, and the cause of death for all those who died. His analysis was an early effort at constructing a life table. Other countries began to pick up on what Graunt was doing, and they, too, began keeping mortality tables. In 1669, Christian Huygens began investigating life expectancy.

Within a generation of Graunt's death, there were early beginnings of the life insurance business. And, by 1693, Edmund Halley had figured out how to create a reliable table of life expectancy for calculating annuities. The early life insurance companies established in London in the 18th century used Halley's table. (These tables were used later to test the efficacy of inoculation against smallpox.)

From a paternalistic but hands-off stance, countries began to take a more active approach in overseeing proper sanitation and health care. Care of orphans, supervision of midwives, discouraging use of tobacco and spirits, and inspection of food and water eventually began to be undertaken by governments. However, for a very long time, it was really theoretical.

SANITATION DURING THESE YEARS

Towns during the 16th and 17th centuries were more like medieval communities than they were like any urban center today. They provided space for a market for the surrounding area, vegetable gardens were within town walls, and cattle frequently grazed within the town as well.

To minimize waste, regulations were created to keep butchers and fishmongers from throwing waste into gutters or into the streams where towns obtained water. There were specific punishments spelled out for those who polluted with human or animal waste. In the 17th century, some towns began to prohibit animals from roaming the streets, and some municipalities set up town privies. But sewage problems were far from solved. If there were gardens within the town, then townspeople could use their waste products as fertilizer, while larger towns would stipulate a few places outside the town where people were to take their waste. However, this required citizen cooperation without much ability to oversee enforcement. Some towns turned to scavengers who were hired to collect the waste, and by the 17th century this method of ridding towns of garbage was becoming more common. Written documentation from Dublin show that they hired an outside contractor to take away the waste, although the company did not necessarily do a very good job.

In many locations, town residents were in charge of street cleaning. In England, they asked for weekly sweepings from citizens. In Coventry and Ipswich, each householder had to clean and sweep the streets in front of his door every Saturday. By the 17th century, Gloucester had adopted this rule, and Cambridge required that all paved streets had to be swept twice a week. Inspectors made rounds regularly to be certain that people complied.

Clean Water

Maintaining a clean water supply was also difficult. Dysentery was common in France and England because of the challenges of guarding against pollution. Wells and springs within a town generally provided water, but the old Roman aqueducts provided water in some communities, and some towns collected water in a central cistern where inhabitants drew water. Before the 17th century, it was rare to have water that ran directly into private homes. (Leeds in the late 17th century was one of the first places to bring water into homes.) Sometimes there was a water shortage, which necessitated rationing. In the summer of 1608, Northampton had a dry summer and had to turn water off at public taps from 10 A.M. to 2 P.M. and again from 7 P.M. to 6 A.M.

London had an abundance of water at first, with the Thames, the Fleet, and the Walbrook Rivers all flowing nearby, but by the end of the Elizabethan era there wasn't enough water and they had to bring it in from outside. In one of the first private solutions in 1609, Sir Hugh Myddleton, a goldsmith and a citizen of London, offered to create an enterprise to bring in water: the New River Company. King James I agreed and Myddleton arranged for water to be brought to Islington Reservoir. These kinds of solutions did not really last for a long time.

In York, they experimented with a purification system. Water was brought from the river in a big pot and then left for a day or two to let the sediment settle. In the 17th century, L. A. Porzio wrote a book about the health of soldiers and suggested using sand to filter the water. On a citywide basis, this was not employed until the 19th century. On a house-by-house basis, French families began to use this method to purify what they drank much earlier.

CARE OF THE SICK

Provision of medical care for the "lame, the halt, and the blind" received lip service, but no one did a very good job of taking care of the infirm or the ill. In England, hospitals quit being a priority after monasteries ended under Henry VIII. If they existed, they were part poorhouse, part home for the aged, and provided very basic care for the sick.

France and Germany also began to turn hospitals over to the government, but the care offered under these circumstances was not very good. In Paris, they started establishing general hospitals that were a combination of poorhouse and hospital. By the 17th century, there were some signs of improvement when hospitals began to be used for teaching medicine. This was a very positive development. One of the first countries to do this was the Netherlands where a teaching hospital was established in Leyden in 1626. Hermann Boerhaave (1668–1738), who strongly believed in this form of education, wrote a very important work *Institutiones medicae in usus annuae exercitationis domesticos digestae* (1708) in support of this trend.

CONCLUSION

While physicians and scientists during the period from 1450 to 1700 were beginning to make important advances in various aspects of science and people such as Sydenham were improving patient care by doing a better job of paying attention to both patient and disease, for the most part the gains were not having a big impact yet on the state of medicine. Sanitation was still not well understood, nor was what was understood well executed, and so clean water and clean streets were still more a matter of luck than good policy. However, while the areas of progress were still not being applied in ways that improved the population's health or improved the prognosis of those who were sick, the foundation was being laid for major moves forward in the future.

CHRONOLOGY

1347–48	The Black Death reaches Europe.
1439	Johannes Gutenberg devised a method of printing using metal molds and alloys to create movable type (printing press).
1452–1519	Leonardo da Vinci drew anatomically accurate drawings of the human body.
1492	Columbus arrives in the Americas.
1490s	Syphilis begins to spread in Europe.
1493–1541	Paracelsus rejected Galen's humoral balance theory and used the principles of alchemy to make medications.
16th century	Anatomy became an important foundation for Western medicine.
1507–87	Thomas Gale, a British surgeon, crusaded against charlatans.
1510–90	Lifetime of Ambroise Paré, who wrote *La méthode de traicter les playes faictes par les hacquebutes et aultres bastons à feu . . .* (Method of treating wounds made by harquebuses and other guns . . .)
1511–53	Miguel Serveto was the first to develop a coherent understanding of pulmonary circulation.
ca. 1516–59	Realdo Colombo, an Italian apothecary, became the first professor of anatomy at the University of Pisa and the first well-known anatomist to write on pulmonary circulation.
1520–74	Bartolomeo Eustachio, considered one of the founders of modern anatomy, discovered the eustachian tubes, the suprarenals, the thoracic duct, and the abducen nerve.

1523–62 Gabriele Falloppio was associated with the discovery of the fallopian tubes, but his primary focus was on the anatomy of the head and ear.

1537 Paré develops a new method of treating gunpowder wounds.

1537 Laws changed permitting autopsies on an as-needed basis.

1540 The Guild of Surgeons merged with the Barbers Company to form the Barber-Surgeons Company.

1543 Nicolaus Copernicus (Mikolaj Kopernik) published *De revolutionibus orbium colestium* (On the revolution of the heavenly spheres).

1543 Andreas Vesalius published *De humani corporis fabrica* (On the fabric of the human body).

1544–1603 William Clowes, master of wound treatment, wrote *A Prooved Practice for all young Chirugeons, concerning Burnings with gunpowder, and woundes made with Gunshot, Sword, Halbard, Pike, Launce or such other* (1588).

1546–99 Gasparo Tagliacozzi revived the art of rhinoplasty.

1561–1636 Santorio created the first method of studying metabolism, a steelyard balance, and a device he called a pulsilogium that measured the pulse.

1563–1636 Louyse Bourgeois, an influential midwife, increased the level of professionalism among those who oversaw the birthing process and published one of the first treatises on midwifery.

1578–1657 William Harvey explained his belief that the blood was circulated by the heart within a closed circulatory system. Harvey believed that all living things originated from an embryo that was found in the egg.

1596–1650	René Descartes wrote *The Description of the Human Body,* in which he suggested that the arteries and veins were pipes that carried nourishment around the body.
1603	Fabricius published *On the Valves of the Veins.*
1621–75	Thomas Willis identified puerperal fever (childbirth fever) and began distinguishing among different forms of diabetes.
ca. 1623–86	Richard Wiseman was considered one of the greatest surgeons of the 17th century.
1627–91	Robert Boyle devised the theory that everything was composed of minute but not indivisible particles of a single universal matter.
1628–94	Marcello Malpighi discovered capillaries, founded the science of microscopic anatomy, and was the first histologist.
1630s	Plants like Peruvian bark, tobacco, and cinchona begin to be imported from the New World and used as medicines.
1649	Nicholas Culpeper takes a stand for the common man and makes a point of trying to treat only the underprivileged.
1660	The Royal Society of London was founded for the "promotion of Natural Knowledge."
1665	*Microphagia* by Robert Hooke is published.
1660s	Thomas Sydenham revives Hippocrates' theory of observation-based medicine.
1670s	William Petty lays the foundation for gathering and evaluating health data quantitatively.
1668	Redi investigates spontaneous generation with red meat and maggots.
1683	Leeuwenhoek sees "little animalcules."
1700	Bernardino Ramazzini publishes his book on occupational illnesses.

GLOSSARY

abducens nerve either of the sixth pair of cranial nerves that are motor nerves supplying the rectus on the outer and lateral side of each eye

abiogenesis the supposed spontaneous origination of living organisms directly from lifeless matter

alchemy a medical chemical science and speculative philosophy aiming to achieve the transmutation of the base metals into gold, the discovery of a universal cure for disease, and the discovery of a means of indefinitely prolonging life

alkaloid any of numerous, usually colorless, complex and bitter organic bases (as morphine or caffeine) containing nitrogen and usually oxygen that occur especially in seed plants and are typically physiologically active

anemia a condition in which the blood is deficient in red blood cells, in hemoglobin, or in total volume

anatomy the act of separating the parts of the organism in order to ascertain their position, relations, structure, and function

anesthetic a substance that produces anesthesia; something that brings relief

antidote a remedy to counteract the effects of poison

antiseptic opposing sepsis, putrefaction, or decay; especially: preventing or arresting the growth of microorganisms

apothecary one who prepares and sells drugs or compounds for medicinal purposes

artery tubular branching muscular- and elastic-walled vessel that carries blood from the heart through the body

astrology the divination of the supposed influences of the stars and planets on human affairs and terrestrial events by their positions and aspects

astronomy the study of objects and matter outside the Earth's atmosphere and of their physical and chemical properties

bryozoan any of a phylum of aquatic mostly marine invertebrate animals

capillaries a capillary tube: especially: any of the smallest blood vessels connecting arterioles with venules and forming networks throughout the body

cautery the act or effect of cauterizing; an agent (as a hot iron or caustic) used to burn, sear, or destroy tissue

chancre a primary sore or ulcer at the site of entry of a pathogen; especially: the initial lesion of syphilis

chilblains an inflammatory swelling or sore caused by exposure to cold

cochlea a hollow tube in the inner ear of higher vertebrates that is usually coiled like a snail shell and contains the sensory organ of hearing

convex curved or rounded outward like the exterior of a sphere or circle

corpuscle a living cell; especially one (as a red or white blood cell) not aggregated into continuous tissues

decoction an extraction gained by boiling down something into a concentrate

diastole a rhythmically recurrent expansion; especially: the dilation of the cavities of the heart during which they fill with blood

dissect to separate into pieces: expose the several parts of for scientific examination

dogmatism positiveness in assertion of opinion, especially when unwarranted or arrogant

dysentery a disease characterized by severe diarrhea with passage of mucus and blood and usually caused by infection

efficacious having the power to product a desired effect

elixir a sweetened liquid usually containing alcohol that is used in medication whether for its medicinal ingredients or as a flavoring

emetic causing vomiting

epidemic affecting or tending to affect a disproportionately large number of individuals within a population, community, or region at the same time

epigenesis development of a plant or animal from an egg or spore through a series of processes in which unorganized cell masses differentiate into organs and organ systems; also: the theory that plant and animal development proceeds in this way

excreta waste matter (as feces) eliminated or separated from the body

expectorant an agent that promotes the discharge or expulsion of mucus from the respiratory tract

fallopian tubes pair of tubes that carry the egg from the ovary to the uterus

foraminifer a type of marine protozoan usually having a shell with a high proportion of calcium

gout a metabolic disease marked by a painful inflammation of the joints, deposits in and around the joints, and usually an excessive amount of uric acid in the blood

harquebus a matchlock gun invented in the 15th century that was portable but heavy and was usually fired from a support

heliocentric referred to or measured from the Sun's center or appearing as if seen from it

hemorrhage a copious discharge of blood from the blood vessels

hemostat clamp an instrument for compressing a bleeding vessel

henbane a poisonous fetid Eurasian herb (*Hyoscyamus niger*) of the nightshade family with yellowish-brown flowers and sticky hairy leaves

histologist a branch of anatomy that deals with the minute structure of animal and plant tissues as discernible with the microscope

iconic a usually pictorial representation; an image; an object of uncritical devotion

iconoclast a person who attacks settled beliefs or institutions

immunity a quality or state of being immune; a condition of being able to resist a particular disease, especially through preventing development of a pathogenic microorganism or by counteracting the effects of its products

indigenous having originated in or being produced, growing, living, or occurring naturally in a particular region or environment

infusion the slow introduction of a solution

inoculate to introduce a microorganism

in utero in the uterus; before birth

jaundice yellowish pigmentation of the skin, tissues, and body fluids caused by the disposition of bile pigments

laticifer a plant cell or vessel that contains latex

laudanum any of various formerly used preparations of opium

ligature something that is used to bind; specifically: a filament used in surgery

membrane a thin soft pliable sheet or layer especially of animal or plant origin

metabolism a sum of the processes in the build up and distribution of protoplasm; specifically: the chemical changes in living cells by which energy is provided for vital processes and activities and new material is assimilated

midwife a person who assists women in childbirth

musket a heavy large-caliber muzzle-loading usually smooth-bore shoulder firearm; broadly: a shoulder gun carried by infantry

obstetrics a branch of medical science that deals with birth and with its antecedents and sequels

opium a bitter brownish addictive narcotic drug that consists of the dried latex obtained from immature seed capsules of the opium poppy

peritoneum the smooth transparent serous membrane that lines the cavity of the abdomen of a mammal and is folded inward over the abdominal and pelvic viscera

pharmacist a person licensed to engage in pharmacy

phlogiston the hypothetical principle of fire regarded formerly as a material substance

physician a person skilled in the act of healing; specifically: one educated, clinically experienced, and licensed to practice medicine as usually distinguished from surgery

physiology a branch of biology that deals with the functions and activities of life or of living matter (as organs, tissues, or cells) and of the physical and chemical phenomena involved—compare anatomy

pleura a delicate serous membrane that lines each half of the thorax of mammals and is folded back over the surface of the lung of the same side

predestination the act of predestinating: the state of being predestined

purgative a medicine causing the removal of undesirable elements

quantification the operation of quantifying (counting)

quicksilver mercury

sacrilege a technical and not necessarily intrinsically outrageous violation (as improper reception of a sacrament) of which is sacred because consecrated to God

scarlet fever an acute contagious febrile disease caused by hemolytic Group A streptococci and characterized by inflammation of the nose, throat, and mouth, generalized toxemia, and a red rash

scurvy a disease caused by a lack of vitamin C and characterized by spongy gums, loosing of the teeth, and a bleeding into the skin and mucous membranes

septum a dividing wall or membrane especially between bodily spaces or masses of soft tissue

shaman priest or priestess who uses magic to cure sickness

sinew tendon

smallpox an acute contagious febrile disease of humans that is caused by a pox virus

spirochete any of the order of slender spirally undulated bacteria including those causing syphilis and Lyme disease

steelyard balance a balance in which an object to be weighed is suspended from the shorter arm of a lever and the weight determined by moving a counterpoise along a graduated scale on the longer arm until equilibrium is attained

suprarenal situated above or anterior to the kidneys

systole a rhythmically recurrent contraction; especially: the contraction of the heart by which the blood is forced onward and the circulation kept up

tendon a tough cord or band of dense white fibrous connective tissue that unites a muscle with some other part (as a bone) and transmits the force which the muscle exerts

thorax the part of the mammal body between the neck and the abdomen; also: the cavity in which the heart and lungs lie

tincture a solution of medicinal substance in an alcoholic solvent

tumor a swollen distended part; an abnormal benign or malignant new growth of tissue that possesses no physiological function and arises from controlled usually rapid cellular proliferation

vesicular containing, composed of, or characterized by vesicles

vivisection the cutting of or operation on a living animal usually for physiological or pathological investigation

FURTHER RESOURCES

ABOUT SCIENCE AND HISTORY

Diamond, Jared. *Guns, Germs, and Steel: The Fates of Human Societies.* New York: W. W. Norton, 1999. Diamond places the development of human society in context, which is vital to understanding the development of medicine.

Dubus, Allen G. *Man and Nature in the Renaissance.* Cambridge: Cambridge University Press, 1978. This book has quotes from Vesalius, which is very helpful in understanding his work.

Hazen, Robert M., and James Trefil. *Science Matters: Achieving Scientific Literacy.* New York: Doubleday, 1991. A clear and readable overview of scientific principles and how they apply in today's world, which includes the world of medicine.

Internet History of Science Sourcebook. Available online. URL: http://www.fordham.edu/halsall/science/sciencsbook.html. Accessed July 9, 2008. A rich resource of links related to every era of science history, broken down by disciplines, and exploring philosophical and ethical issues relevant to science and science history.

Lindberg, David C. *The Beginnings of Western Science.* 2nd ed. Chicago: University of Chicago Press, 2007. A helpful explanation of the beginning of science and scientific thought. Though the emphasis is on science in general, there is a chapter on Greek and Roman medicine as well as medicine in medieval times.

Roberts, J. M. *A Short History of the World.* Oxford: Oxford University Press, 1993. This helps place medical developments in context with world events.

Silver, Brian L. *The Ascent of Science.* New York: Oxford University Press, 1998. A sweeping overview of the history of science from the Renaissance to the present.

Spangenburg, Ray, and Diane Kit Moser. *The Birth of Science: Ancient Times to 1699,* rev. ed. New York: Facts On File, 2004. A highly readable book with key chapters on some of the most significant developments in medicine.

ABOUT THE HISTORY OF MEDICINE

Ackerknecht, Erwin H., M.D. *A Short History of Medicine,* rev. ed. Baltimore: Johns Hopkins University, 1968. While there have been many new discoveries since Ackerknecht last updated this book, his contributions are still important as it helps the modern researcher better understand when certain discoveries were made and how viewpoints have changed over time.

Arano, Luisa Cogliati. *The Medieval Health Handbook,* New York: George Braziller, 1976. This was originally published in Italy as *Tacuinum Sanitatis* and describes medieval cures of the day.

Bishop, W. J. *The Early History of Surgery.* London: The Scientific Book Guild, 1960. This book is dated but helpful on the history of surgery.

Clendening, Logan, ed. *Source Book of Medical History.* New York: Dover Publications, 1942. Clendening has collected excerpts from medical writings from as early as the time of the Egyptian papyri, making this a very valuable reference work.

Dary, David. *Frontier Medicine: From the Atlantic to the Pacific 1492–1941.* New York: Alfred A. Knopf, 2008. This is a brand-new book that has been very well reviewed. Dary outlines the medical practices in the United States from 1492 on.

Davies, Gill, ed. *Timetables of Medicine.* New York: Black Dog & Leventhal, 2000. An easy-to-assess chart/time line of medicine with overviews of each period and sidebars on key people and developments in medicine.

Dawson, Ian. *The History of Medicine: Medicine in the Middle Ages.* New York: Enchanted Lion Books, 2005. A heavily illustrated short book to introduce young people to what medicine was like during medieval times. Dawson is British so there is additional detail about the development of medicine in Britain.

Dittrick Medical History Center at Case Western Reserve. Available online. URL: http://www.cwru.edu/artsci/dittrick/site2/. This site provides helpful links to medical museum Web sites. Accessed October 31, 2008.

Duffin, Jacalyn. *History of Medicine.* Toronto: University of Toronto Press, 1999. Though the book is written by only one author, each chapter focuses on the history of a single aspect of medicine, such as surgery or pharmacology. It is a helpful reference book.

Haeger, Knut. *The Illustrated History of Surgery.* Gothenburg, Sweden: AB Nordbok, 1988. This is an academic book that is very helpful in understanding early surgery.

Kennedy, Michael T., M.D., FACS. *A Brief History of Disease, Science, and Medicine.* Mission Viejo, Cal.: Asklepiad Press, 2004. Michael Kennedy was a vascular surgeon and now teaches first- and second-year medical students an introduction to clinical medicine at the University of Southern California. The book started as a series of his lectures, but he has woven the material together to offer a cohesive overview of medicine.

Loudon, Irvine, ed. *Western Medicine: An Illustrated History.* Oxford: Oxford University Press, 1997. A variety of experts contribute chapters to this book that covers medicine from Hippocrates through the 20th century.

Magner, Lois N. *A History of Medicine.* Boca Raton, Fla.: Taylor & Francis Group, 2005. An excellent overview of the world of medicine from paleopathology to microbiology.

Porter, Roy, ed. *The Cambridge Illustrated History of Medicine.* Cambridge, Mass.: Cambridge University Press, 2001. In essays written by experts in the field, this illustrated history traces the evolution of medicine from the contributions made by early Greek physicians through the Renaissance, Scientific Revolution, and 19th and 20th centuries up to current advances. Sidebars cover parallel social or political events and certain diseases.

Porter, Roy. *The Greatest Benefit to Mankind: A Medical History of Humanity.* New York: W. W. Norton, 1997. Over his lifetime, Porter wrote a great amount about the history of medicine, and this book is a valuable and readable detailed description of the history of medicine.

Rosen, George. *A History of Public Health, Expanded Edition.* Baltimore: Johns Hopkins University Press, 1993. While serious public health programs did not get underway until the 19th century, Rosen begins with some of the successes and failures of much earlier times.

Simmons, John Galbraith. *Doctors & Discoveries.* Boston: Houghton Mifflin Company, 2002. This book focuses on the personalities behind the discoveries and adds a human dimension to the history of medicine.

Toledo-Pereyra, Luis H. *A History of American Medicine from the Colonial Period to the Early Twentieth Century.* Lewiston, N.Y.: Edwin Mellen Press, 2006. This is an academic book that provides very valuable information about colonial America.

United States National Library, National Institutes of Health. Available online. URL: http://www.nlm.nih.gov/hmd/. Accessed July 10, 2008. A reliable resource for online information pertaining to the history of medicine.

OTHER RESOURCES

Annenberg Media Learner.org. Available online. URL: http://www.learner.org/interactives/middleages/morhealt.html. Accessed October 31, 2008. Information on medieval medicine with links to other medieval sites.

Ford, Brian J., and Al Shinn. "Antoni van Leeuwenhoek (1632–1723)." University of California at Berkeley Museum of Paleontology Web site. Available online. URL: http://www.ucmp.berkeley.edu/history/leeuwenhoek.html. Accessed December 3, 2008. An incisive essay on the scientist's research from the point of view of his use of the microscope.

Newman, Paul B. *Daily Life in the Middle Ages.* Jefferson, N.C.: McFarland & Company, 2001. This is a wonderfully thorough book about life in the middle ages, and it describes everything from what they ate to how they fought during medieval times.

Ryves, W. "The life of the admired physician and astrologer of our times, Mr. Nicholas Culpeper." Published in *Culpeper's School of Physick, 1659.* Available online. URL: http://www.skyscript.co.uk/culpeper.html. Accessed December 15, 2008. Culpeper was a fascinating fellow.

Sacks, Oliver. *Migraine.* New York: Vintage Press, 1999. A helpful book about understanding migraines.

Smith, Alan DeForest. "Richard Wiseman: His Contributions to English Surgery." In *Bulletin of the New York Academy of Medicine* 46, no. 3 (March 1970). This provides information on Wiseman and surgery.

INDEX

Note: Page numbers in *italic* refer to illustrations; *m* indicates a map; *t* indicates a table.

H

I

J

K

L

M

Q

R

S